Smartphone Based Medical Diagnostics

Smartphone Based Medical Diagnostics

Edited by

Jeong-Yeol Yoon, PhD

Professor
Department of Biomedical Engineering
Department of Biosystems Engineering
The University of Arizona
Tucson, Arizona
United States

ACADEMIC PRESS

An imprint of Elsevier

Academic Press is an imprint of Elsevier
125 London Wall, London EC2Y 5AS, United Kingdom
525 B Street, Suite 1650, San Diego, CA 92101, United States
50 Hampshire Street, 5th Floor, Cambridge, MA 02139, United States
The Boulevard, Langford Lane, Kidlington, Oxford OX5 1GB, United Kingdom

Smartphone Based Medical Diagnostics

Notices
Practitioners and researchers must always rely on their own experience and knowledge in
evaluating and using any information, methods, compounds or experiments described
herein. Because of rapid advances in the medical sciences, in particular, independent
verification of diagnoses and drug dosages should be made. To the fullest extent of the
law, no responsibility is assumed by Elsevier, authors, editors or contributors for any injury
and/or damage to persons or property as a matter of products liability, negligence or
otherwise, or from any use or operation of any methods, products, instructions, or ideas
contained in the material herein.

ISBN: 978-0-12-817044-1

Publisher: Mara Conner
Acquisition Editor: Fiona Geraghty
Editorial Project Manager: Fernanda A. Oliveira
Production Project Manager: Sreejith Viswanathan
Cover Designer: Alan Studholme

Contents

Contributors

Jokubas Ausra, BS
Department of Biomedical Engineering, University of Arizona, Tucson, AZ, United States

Alex Burton, BS
Department of Biomedical Engineering, University of Arizona, Tucson, AZ, United States

Cheng Gong
College of Optical Sciences, The University of Arizona, Tucson, AZ, United States

Philipp Gutruf, BS, PhD
Assistant Professor, Department of Biomedical Engineering, BIO5 Institute, Department of Electrical Engineering, University of Arizona, Tucson, AZ, United States

Yushin Ha, PhD
Professor, Bio-Industrial Machinery Engineering, Kyungpook National University, Daegu, Republic of Korea

Kwan Young Jeong
Department of Applied Chemistry & Biological Engineering and Department of Molecular Science & Technology, Ajou University, Suwon, Republic of Korea

Dongkyun Kang, PhD
Assistant Professor, College of Optical Sciences, University of Arizona, Tucson, AZ, United States

Dong Woo Kim
Department of Applied Chemistry & Biological Engineering and Department of Molecular Science & Technology, Ajou University, Suwon, Republic of Korea

Nachiket Kulkarni
College of Optical Sciences, The University of Arizona, Tucson, AZ, United States

Zheng Li, PhD
Department of Chemical and Biomolecular Engineering, North Carolina State University, Raleigh, NC, United States

Bijan Najafi, PhD, MSc
Professor of Surgery, Interdisciplinary Consortium for Advanced Motion Performance (iCAMP), Division of Vascular Surgery and Endovascular Therapy, Michael E. DeBakey Department of Surgery, Baylor College of Medicine, Houston, TX, United States

Christopher David Nguyen
College of Optical Sciences, The University of Arizona, Tucson, AZ, United States

Tusan Park, PhD
Professor, Bio-Industrial Machinery Engineering, Kyungpook National University, Daegu, Republic of Korea

Anna Pyayt, PhD
Associate Professor, Chemical & Biomedical Engineering, University of South Florida, Tampa, FL, United States

Wonjin Shin
Department of Bio-Industrial Machinery Engineering, Kyungpook National University, Daegu, Republic of Korea

Tucker Stuart, BS
Department of Biomedical Engineering, University of Arizona, Tucson, AZ, United States

Daniel Dooyum Uyeh, PhD
Department of Bio-Industrial Machinery Engineering, Kyungpook National University, Daegu, Republic of Korea

Qingshan Wei, PhD
Assistant Professor, Department of Chemical and Biomolecular Engineering, North Carolina State University, Raleigh, NC, United States

Hyun C. Yoon, PhD
Professor, Department of Applied Chemistry and Biological Engineering, Department of Molecular Science and Technology, Ajou University, Suwon, Republic of Korea

Jeong-Yeol Yoon, PhD
Professor, Department of Biomedical Engineering and Department of Biosystems Engineering, The University of Arizona, Tucson, AZ, United States

Han Zhang
Department of Biological Engineering, Utah State University, Logan, UT, United States

Shengwei Zhang
Department of Chemical and Biomolecular Engineering, North Carolina State University, Raleigh, NC, United States

Wei Zhang
Department of Biological Engineering, Utah State University, Logan, UT, United States

Anhong Zhou, PhD
Professor, Department of Biological Engineering, Utah State University, Logan, UT, United States

Wenbin Zhu, PhD
College of Optical Sciences, The University of Arizona, Tucson, AZ, United States

Introduction

Jeong-Yeol Yoon, PhD

*Professor, Department of Biomedical Engineering and Department of Biosystems Engineering,
The University of Arizona, Tucson, AZ, United States*

Overview of the entire book is addressed in this chapter. A couple of basic terms are defined, including biomedical diagnostics and biosensors. A basic concept of biosensors is then introduced, and two most popular transducers—optical and electrochemical—are briefly explained. The sensors and connectivity options available in many commercial smartphones are explained, and their utilization toward biosensing and ultimately biomedical diagnostics is explained. A brief introduction on the remaining chapters of this book is summarized at the end of this chapter.

1. Some definitions: medical diagnostics and biosensors

The title of this textbook is "Smartphone Based Medical Diagnostics." *Diagnostics*, by definition, is the discipline (like mathematics) of diagnosis. Diagnostics is sometimes abbreviated *Dx*. *Diagnosis*, by definition, is the identification of the nature and/or cause of a certain phenomenon. It is sometimes abbreviated *Ds*. In *medical diagnostics*, the phenomenon is typically a disease, disorder, and/or syndrome in a human subject. The nature and cause of disease, disorder, or syndrome vary substantially, ranging from pathogens (including bacteria, viruses, and fungi), toxic chemicals, cancer, genetic disorder, and traumatic injury, to name a few. Medical diagnostics can be expanded to *biomedical diagnostics*, to include the phenomena occurring in nonhuman subjects, such as plants, animals, food, water, and air, all of which can eventually affect human health.

In a traditional sense, medical and biomedical diagnostics should be performed in a wet laboratory, where skilled and trained personnel conduct a series of sample (mostly liquid) handling procedures using a variety of analytical instruments. Such laboratory-based medical diagnoses have been replaced with standalone *biosensors* in the past couple of decades, greatly reducing the cost and time of such diagnoses, as well as allowing the diagnosis to be performed not in a remote laboratory, but at the *point of care* (called as *POC diagnostics* or *POC Dx*), or even at the convenience of patient's own home [1].

Smartphone Based Medical Diagnostics. https://doi.org/10.1016/B978-0-12-817044-1.00001-6

FIGURE 1.1

Three most popular commercial biosensors. Top left: a glucose meter (with a glucose strip inserted) is measuring blood glucose concentration from a finger prick blood sample. Top right: a pregnancy test is evaluating a pregnancy hormone (hCG) from urine sample. Bottom: a pulse oximeter is measuring pulse and blood oxygen saturation from a human finger.

Top left: Reprinted from Ref. [1] with permission, (C) 2016 Springer. Top right: Reprinted from Ref. [2]. Bottom: Reprinted from Ref. [3].

At the time of writing, three biosensors have been most successfully commercialized and are widely used at hospitals, doctor's offices, and homes—*glucose meter, pregnancy test*, and *pulse oximeter* (Fig. 1.1) [1−3]. In all three cases, the biosensors are detecting targets whose concentrations are very high—glucose in blood, pregnancy hormone (human chorionic gonadotropin or hCG) in urine, and hemoglobin in blood—and can be detected relatively easily. Of course, many other biosensors are also available commercially and new types of biosensors are continuously emerging at this time, often targeting the molecules whose concentrations are substantially lower than those listed above.

2. What is biosensor?

Biosensor is one type of sensor that can identify the type/species and/or the concentration of biological analyte. Examples of bioanalytes include a simple biochemical compound (e.g., glucose), a sequence of nucleic acid (DNA or RNA), a specific protein, a virus particle, a bacterium, and so on [1]. The presence of these biological analytes and/or their concentrations can then be utilized to identify the nature and cause of disease, disorder, or syndrome. Therefore, biosensors are mostly used for biomedical diagnostics.

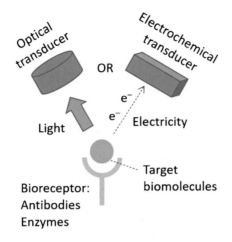

FIGURE 1.2

A typical biosensor.

To identify and quantify these bioanalytes, *bioreceptors* are necessary (Fig. 1.2). Bioreceptors bind to the target analytes in a highly specific manner. Obviously, a wide variety of bioreceptors have been tested and evaluated for biosensors and subsequently medical diagnostics. The following two types of bioreceptors have been used most frequently: (1) antibodies and (2) enzymes. *Antibodies* are protein molecules normally found in human (as well as animal) blood. They bind to the target antigens and thus nullify the antigen's action in the body—hence they form a part of human's (as well as animal's) immune system. They are relatively specific, for example, antibody to the well-known bacterium *Escherichia coli* (i.e., anti-*E. coli*) binds only to *E. coli* but not to other bacteria, viruses, or proteins. (In reality, anti-*E. coli* can also bind to the other bacteria that are similar to *E. coli*, called as cross-binding; however, such probability is relatively low and its specificity is still substantially superior to other chemical ligands.) The target antigens are typically proteins. When antibodies bind to bacteria or viruses, they actually bind to their surface proteins. *Enzymes* are also protein molecules found in human (or animal) bodies. They bind to the target substrate and catalyze a chemical reaction (most commonly oxidation) of that substrate. Therefore, enzymes are often called as biological catalysts. Substrates are typically small chemical compounds, for example, glucose, cholesterol, alcohol, etc. In addition, enzymes are relatively specific to the target substrate, similar to antibodies.

Once bioreceptor specifically binds to the target bioanalytes, it becomes necessary to quantify the extent of such binding. Such quantifications are performed by transducers (Fig. 1.2). *Transducer*, in fact, is a pivotal component not just in biosensors but also in all sensors, where the type and concentration of bioanalytes (in biosensors) or the physical property (in sensors) are converted into analog voltage signals, which are further converted to digital signals.

Although other types of transducers are available for biosensors (e.g., piezoelectric and thermal), the following two types are the most commonly used: optical and electrochemical. Biosensors with optical transducers are typically called as *optical biosensors* and those with electrochemical transducers as *electrochemical biosensors*. Both types of biosensors are explained in Chapter 2 and Chapter 3.

Both optical and electrochemical biosensing can be performed using analytical instruments in a wet laboratory. Optical biosensing is conducted typically using a spectrophotometer, and electrochemical biosensing using electrodes (e.g., pH, ion-selective, or conductivity electrode) or an impedance analyzer. Commercial optical or electrochemical biosensors are typically simplified versions of such analytical instruments, tailored for a specific application. The major downside of this approach is that each application needs a specific biosensor device, while analytical instruments can be used for multiple (or even general) applications.

As implied in the title of this book, it is also possible to conduct optical and electrochemical biosensing using a smartphone. Considering the widespread availability of smartphones, this approach will certainly reduce the effort and cost of developing a specific biosensor, reduce the actual cost of assays, and allow the general public to familiarize themselves to new biosensor technology. In addition, smartphones carry advanced processing units (their computing power far outperforming those incorporated in commercial biosensors), a large amount of memory for data storage, and ability to send the raw and processed assay results to a cloud storage and/or other mobile device. These features are not possible with commercial biosensors, unless a separate laptop computer (which severely compromises the portability of the biosensor) is connected to them.

3. What sensors are available in smartphones for biosensing?

Before the era of smartphone, cellular phones (or mobile phones) were also equipped with a couple of extra features other than voice calls, including text messaging and limited data networking capabilities. To make a distinction from modern smartphones, such cellular (or mobile) phones are retroactively called as *feature phones*. In fact, digital data transmission has made possible with *2G* (*second generation*) cellular technology, where voice, text, and data are all transmitted in digitally encrypted fashion. (In comparison, the 1G cellular technology was based on analog data transmission.) With digital transmission, it became possible to transmit multimedia text messages, such as photographs. Therefore, most 2G feature phones have incorporated a small digital camera, which have become enormously popular among users. It was also possible to send emails with photograph attachments via 2G. Unfortunately, those feature phones were generally incapable of WiFi and Bluetooth.

As 2G data transmission was notoriously slow, it was not possible to send high-resolution photographs. Therefore, the camera resolutions of most feature phones were quite low, for example, 640×480 pixels $= 0.3$ *megapixels* (*MP*).

The first iPhone, released in 2007, was revolutionary in many aspects, incorporating a true operating system (OS), substantial memory (dynamic random-access memory or DRAM), flash memory storage (in lieu of a hard disk drive), bigger display, and higher resolution camera (2.0 MP). It was also capable of WiFi and Bluetooth connectivity. However, its major limitation was that it was still based on 2G technology (they advertised it as 2.5G), which was still insufficient to handle internet connectivity (and should have been complemented by WiFi).

One year later, iPhone 3G was introduced, obviously utilizing 3G technology with significantly improved data transmission rate. This is the smartphone that brought Apple Inc. substantial sales revenue and made smartphones widely used by the general public. Google Inc. also released Android OS, an open source and open architecture operating system, specifically for smartphone, and many other companies started producing smartphones (most notably Samsung's Galaxy series).

As smartphones have evolved, more sensors have been incorporated into them (Fig. 1.3). Digital cameras are now equipped not just in rear but also in front. Their resolutions are now comparable to most standalone digital cameras, effectively eradicating their low- to medium-end markets. Rear cameras now come in two or even three (dual or triple cameras), for the sake of improving resolution with better focusing capability. *White LED flash* is also equipped, to provide flashlight. *Ambient light sensor* has later been added to the front side to sense the amount of ambient light, so that the smartphone's screen brightness can be appropriately adjusted automatically for the user's convenience as well as saving the battery. *Proximity sensor* is also essential, which detects the presence of a human ear. When the user places the smartphone to his/her ear, proximity sensor recognizes it, and turns off the display to save the battery and avoid making the phone too hot. It is essentially an IR sensor. Recently, *IR camera* is being added to the smartphone, which is quite useful for *night vision* as well as *thermal imaging*.

Inside the smartphone, it holds additional sensors. *GPS* (*global positioning system*) is quite useful in using maps within smartphone and for navigation applications. *Accelerometer* measures acceleration (including acceleration due to gravity), informing the system whether the smartphone (and subsequently its user) is currently moving or stationary. This feature is quite useful for tracking the user's activity and also an essential feature in *activity trackers* (e.g., Fitbit) as well as *smartwatches*. Accelerometers are piezoelectric devices, where the frequency of piezoelectric material's (e.g., quartz) vibration is measured to relate it to the acceleration. *Gyroscope* measures the orientation and angular velocity. Its most obvious use is the compass app. Using both accelerometer and gyroscope, the smartphone's exact positioning and orientation can be obtained. Although its most obvious use is the automatic screen rotation feature, it can also be used for games, where smartphone's positioning replaces the function of gamepad or joystick. Nintendo's Wii is a good example of utilizing accelerometer and gyroscope as a game controller.

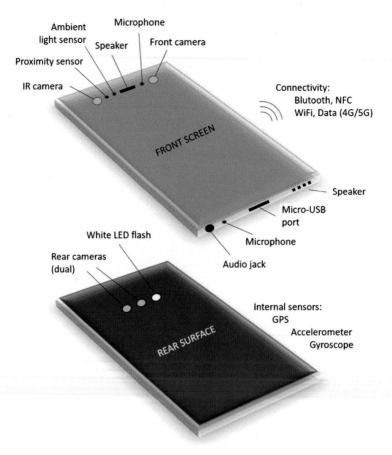

FIGURE 1.3

Various sensors and connectivity options available in smartphones.

The use of accelerometer and gyroscope in smartphone as well as smartwatches is particularly important in evaluating and tracking the patient's activity of physical therapy, and the example for monitoring diabetic foot ulcer is described in Chapter 8.

Many of these sensors are optical sensors—front camera, rear camera, ambient light sensor, proximity sensor, IR camera, and white LED flash, which can be utilized for optical biosensing. The use of smartphone toward optical biosensing is further explained in Chapter 2.

Although smartphones can process, display, and save the data from these internal sensors by itself, it is still important to share these results with other devices. Smartphones can be paired with other devices over short distance via *Bluetooth* (described in Chapter 11 Section 4). Recently, *near field communication* (*NFC*) is also becoming a standard, which is commonly used for mobile payment services. Use of NFC for biomedical diagnostics is briefly described toward microfluidic devices in Chapter 7, and extensively described to be used in conjunction with flexible electronics sensor in Chapter 10. *WiFi* and data network (*4G/LTE* is most commonly used while *5G* is being introduced at the time of writing) allow sending the raw and processed data over internet, to the other smartphones, laptop/desktop computers, and cloud storage. Wired connections are also possible, most importantly a *micro-USB port* (the smallest version of USB connection). The international standard of type C is used for Android phones, while a slightly different version (lightning connector) is used for iPhones. As the micro-USB port can transmit digital data as well as analog voltage, it has a dual purpose—(1) to charge the smartphone's battery from an AC outlet via an AC/DC converter, and (2) to upload or download digital data to and from the smartphone. Both features are quite important in electrochemical biosensing, as it requires an analog voltage applied to the sensor, and collection of voltage/current/resistance signals. Unfortunately, the latter sensor signals are analog, and must be converted to digital signal using a separate *analog-to-digital converter (A/D converter)*. An *audio jack* may serve as a good alternative, as it sends analog sound signals to the headphones and receives analog sound signals from the microphones. A/D converters are incorporated next to the audio jack within smartphones. Unfortunately, this audio jack has been removed from some newer versions of smartphones. The use of smartphone toward electrochemical biosensing is further explained in Chapter 3, and their applications in Chapter 4, Chapter 7 and Chapter 11.

---

4. Overview of chapters

As already described, basic principles of optical and electrochemical biosensing are explained in Chapter 2 and Chapter 3, along with how the sensors and connectivity options in smartphones can be utilized toward such biosensing.

After these theoretical and background explanations, applications of smartphones toward three most common biosensors—glucose meter, pregnancy test, and pulse oximeter—are covered. In Chapter 4, use of smartphone toward glucose sensing is explained, which is one of the very first demonstrations of smartphone-based medical diagnostics. Both optical and electrochemical biosensing methods are explained. Another early example of smartphone-based medical diagnostics is the use of smartphone as an optical sensor device for flow cytometry, covered in Chapter 5.

The next application is a pregnancy test. In fact, pregnancy test is one example of rapid kits, or more precisely, lateral flow immunochromatographic assays or lateral flow assays (LFAs). Normally, they are yes-or-no assays, that is, whether the target (e.g., pregnancy hormone) is present in the sample (e.g., human urine) or not. This method has been expanded to quantify many different hormones, proteins, viruses, and even bacteria from myriads of biological samples, including urine, blood serum, whole blood, saliva, feces (dissolved in buffer), water, food (dissolved in buffer), aerosols (collected by an air sampler), etc. Although rapid kit assays can be conducted on paper strips (e.g., LFAs), they can also be conducted in test tubes. Regardless of its type, it becomes necessary to quantify the coloration intensities from rapid kits to relate them to the target concentration. Although such quantification had previously been conducted using a benchtop apparatus, for example, spectrophotometer or reflectometer, it can easily be achieved by using a smartphone as an optical sensor device. Smartphone-based rapid kits are described in Chapter 6 and smartphone-based LFAs in Chapter 7. LFAs can easily be multiplexed and further improved by conducting the assays on microfluidic devices, also described in Chapter 7. Similar to rapid kits, smartphones can also be used as an optical sensor for microfluidic devices. Smartphone-based electrochemical biosensing from microfluidic devices is also described in that chapter.

Imaging-based optical biosensing is also possible with smartphone, thanks to the recent advancement in its camera and data processing capability. The first example is the smartphone-based monitoring of wounds, especially for diabetic patients. This is described in Chapter 8. Microscopic imaging is also possible, through attaching simple optical accessories to a smartphone. This is described in Chapter 9.

The final application is a pulse oximeter. The pulse meter portion has already been incorporated in many activity trackers and smartwatches. In fact, with the recent advancement of wearables and flexible electronics, smartphone-based biosensors can do more than pulse oximetry, including sensing body temperature, skin pH, and even *electrocardiogram*. This is explained in Chapter 10.

Chapter 11 is devoted to food safety application. Although previous chapters were mostly focused on human health applications, smartphone-based monitoring of food safety has recently emerged, which deserves more attention and further research/development.

5. Regulatory issues

To use these smartphone-based biosensors toward medical diagnostics, it will be necessary to obtain approvals from the government or governmental agency (in the United States, it is *Food and Drug Administration*). Most likely, smartphone should be an integral part of such a biosensor, and its specification must be submitted together for regulatory approval. The problem is that most smartphone manufacturers release new models on an annual basis, while such regulatory approvals take several years. Such delays in regulatory approvals are necessary to ensure its reliability, data privacy, and appropriate data interpretation. Additionally, the smartphone's manufacturer should also share key specification information on their smartphones to the biosensor manufacturer toward successful regulatory approval, which may not be possible in some occasions.

Smartphone-based medical diagnostics could also bring in potential legal and ethical issues. Traditional healthcare law may not be sufficient to address such issues, which may further delay its approval and subsequent marketability. It is important that these issues be brought up early in the development stage and the developers must consult with healthcare providers as early as possible to prevent any further complication and delay in commercialization.

These issues are extensively addressed at the end of Chapter 4, Chapter 5 and Chapter 11.

References

[1] Yoon J-Y. Introduction to biosensors: from electric circuits to immunosensors. 2nd ed. New York, NY: Springer; 2016.
[2] Palma E. Woman with pregnancy test. 2010. Available at: https://commons.wikimedia. org/wiki/File:Dos_rayitas.jpg.
[3] Rcpbasheer. Pulse oximeter. 2007. Available at: https://commons.wikimedia.org/wiki/ File:Puls_oxymeter.jpg.

Basic principles of optical biosensing using a smartphone

Jeong-Yeol Yoon, PhD

*Professor, Department of Biomedical Engineering and Department of Biosystems Engineering,
The University of Arizona, Tucson, AZ, United States*

In the previous chapter, we learned the definition of biosensor and two most popularly used transducers for biosensors: optical and electrochemical. In this chapter, we will learn about the biosensors with optical transducers, or simply optical biosensors, and how such optical biosensing can be demonstrated using a smartphone.

1. Light

As explained in the previous chapter, the type and/or concentration of biochemical, protein, virus, bacterium, or tissue are identified using an optical transducer in optical biosensors. An optical transducer is typically a light sensor. *Light*, in a very narrow sense, is a part of electromagnetic radiation that is visible to human eye. Such *visible light* ranges roughly from 400 nm to 750 nm wavelength. As human eye is not a spectrophotometer, it is unable to recognize and collect a spectrum of such visible light. Instead, human eye has three types of *cone cells* at its innermost part of *retina*, each corresponding to a different "range" of the visible light spectrum: namely red (R), green (G), and blue (B), or RGB, called as trichromatic vision [1]. The response curves of these three different human cone cells are depicted in Fig. 2.1.

As shown in Fig. 2.1, the visible light from 400 nm to 500 nm is mostly picked up by blue cone cells, thus it appears blue; the light from 500 nm to 550 nm is picked up by green cone cells and appears green; and the light from 550 nm 750 nm is picked up by red cone cells and appears red. Blue cone cells are not very sensitive from 400 nm to 430 nm, which should theoretically cause these wavelengths to appear dimly blue. Meanwhile, the peaks of green and red cone cells' response functions are quite close to each other and overlapping, thus the light from 540 to 570 nm is picked up by both green and red cone cells and thus appear yellow (= green + red). In addition, red cone cells are not very sensitive from 650 nm to 750 nm, which cause these wavelengths to appear dimly red.

Different from humans, most vertebrate animals (e.g., dogs) have dichromatic vision, recognizing only two types of colors—blue and yellow. Scientists believe

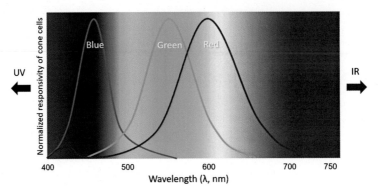

FIGURE 2.1

Colors of visible light shown with their respective wavelength (λ). Normalized responsivities of three different cone cells in human eye's retina are superimposed on it. Note that the peaks of green and red responsivity curves are close to each other. As a result, the peak of the red responsivity curve (around 580 nm) is recognized by both green and red cone cells, resulting in a yellow coloration.

that humans have evolved from dichromatic to trichromatic vision, although the distinction between green and red color is not yet substantial. Some humans are born with difficulty in differentiating green and red color, called as color weakness (weak distinction) and dichromatism (almost no distinction). Some other humans (very rare) are born with four different types of cone cells, recognizing yellow color as the fourth distinctive color in addition to RGB.

Electromagnetic radiation consists of the following based on its wavelength: (1) γ (gamma) ray (<10 pm), (2) X-ray (0.01−10 nm), (3) *ultraviolet* (*UV*) light (10−400 nm), (4) visible light (400−750 nm), (5) *infrared* (*IR*) light (0.8−300 μm), (6) microwave (1 mm−1 m), and (7) radio wave (>1 m). Although all other radiation types are called as "rays" or "waves", UV, visible, and IR radiation are called as "light." In fact, this is the more widely accepted definition of light, incorporating not only visible light but also the ones right next to it. Therefore, a new need has emerged to make a distinction between visible light (often abbreviated *Vis*) and invisible light (UV and IR).

You may wonder why the shorter-than-blue color is named "ultra" and "violet." The shortest wavelength in the visible spectrum, that is, near 400 nm, is mostly recognized by blue cone cells, despite being not very sensitive. The red cone cells, on the other hand, also respond to this wavelength to a smaller extent. Thus, the light at 400 nm and nearby appears purple (= blue + red). However, this monochromatic color is different from the true mixture of blue and red color, and it is called as *violet*. Ultra means "extreme," indicating something big and strong, and in this case is referring to the fact that light at these wavelengths are more energetic than those at the violet wavelengths. This is because shorter wavelengths are correlated to higher energy, as $E = hc/\lambda$, where E is the energy of light, h is the Planck's constant,

c is the speed of light, and λ is the wavelength of light. Accordingly, the longer-than-red color is named infrared, where "infra" means "below," that is, smaller energy than red light.

2. Digital cameras in smartphones

All smartphones (as well as most cellular phones) are equipped with digital cameras. Before the era of smartphone, there were standalone digital cameras, often with a price tag comparable to that of cellular phones. Previously, digital cameras within cellular phones were considered quite inferior versions (in both resolution and sensitivity) of standalone digital cameras. Early versions of smartphones were not very different from cellular phones in terms of their camera's quality. Nowadays, digital cameras within smartphone are so advanced that they have nearly killed the majority market of digital cameras, leaving only high-end, professional digital cameras (e.g., DSLR cameras).

There are many advantages to using digital cameras incorporated within smartphone over standalone digital cameras:

1. Small and lightweight.
2. Cost-effective, as the camera comes with a smartphone.
3. Images and movies can easily be uploaded to cloud storage, emailed, and shared with others, using a smartphone's Wi-Fi or cellular data networking.
4. Images can be processed within a smartphone, using an appropriate application (commercially available or custom coded).

Digital camera is essentially an array of *photodiodes*. Photodiode, by definition, is a special type of diode, where two different types of semiconductors are joined together. One is a *p-type semiconductor*, which is semiconductor material (typically silicon—Si) added with a dopant (typically boron—B) that causes the material to carry a positive charge, thus p-type. The other is an *n-type semiconductor*, added with a dopant (typically phosphorous—P) that causes the material to carry a negative charge, thus n-type. A diode can be made by sandwiching these p-type and n-type semiconductors together. Although diodes can be used for various electric and electronic applications, photodiodes are almost exclusively used to detect light. Fig. 2.2 shows the layout of a typical photodiode [1].

As the p-type semiconductor is positively charged, it contains *holes* (or lack of electrons). Meanwhile, as the n-type semiconductor is negatively charged, it contains free *electrons*. As holes and free electrons are pulled away from each other, a small depletion region is formed at the P-N junction, that is, free of holes and free electrons, and thus not very conductive. If a reverse-bias voltage is applied to this system, holes are attracted more to the negative ($-$) voltage and free electrons to the positive ($+$) voltage, making the depletion region more apparent. Photodiodes can be used both with and without a reverse-bias voltage, called as *photoconductive*

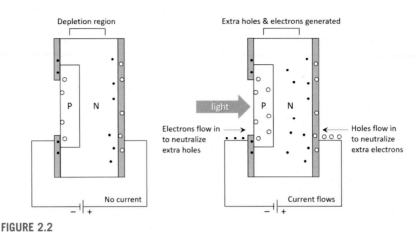

FIGURE 2.2

Photodiode in photoconductive mode.

and *photovoltaic* modes, respectively [1]. Photoconductive mode provides improved sensitivity and is thus more common.

When light hits the built-in window and subsequently the p-type semiconductor, it strips the valence electrons off from the molecules, creating extra holes and free electrons. As the p-type semiconductor is deliberately made very thin, light is also able to hit the top layer of the n-type semiconductor, again creating extra holes and free electrons. Therefore, the depletion region is now full of holes and free electrons, and therefore becomes very conductive. This is a higher energy state and thus unstable. In photovoltaic mode, the p-type and n-type semiconductors are simply connected together with a wire without any voltage source. Extra holes and free electrons simply flow through the wire, eventually merged together, and disappear. This process generates an electric current. This is also a working principle of a *solar cell*.

In photoconductive mode, extra holes and free electrons are neutralized by the incoming electrons and holes from the voltage source. With no light, no current flows due to the presence of the depletion region. With light, current starts to flow. Applying higher reverse-bias voltage makes the depletion region thicker, indicating more hole—electron pairs can be generated upon light exposure, that is, generating much higher current and making its dynamic range broader.

Although a single photodiode can generate only a single signal—electric current—that can be translated into light intensity, it is possible to make a 2D array of photodiodes to acquire an image. Each array component represents a single photodiode, which is called as *photosite*. A signal from a single photosite provides data for a single pixel in the final acquired image. Typically, there are millions of photosites in such 2D arrays and subsequently millions of pixels (e.g., 10 megapixels or 10 MP) in the acquired images. Previously, *charge-coupled device (CCD) arrays* were used to acquire such images. These days, *complementary metal oxide semiconductor (CMOS) arrays* are more commonly used. Fig. 2.3 shows the basic working principle of CCD and CMOS arrays.

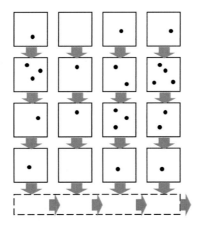

FIGURE 2.3

Working principle of CCD and CMOS arrays.

To make our explanation simpler, a 4×4 array (16 pixel) is shown in Fig. 2.3. In each photosite (photodiode), electrons (charges) are generated at varied amounts, corresponding to an image. These electrons are sequentially transferred to the neighboring photosites and collected at the bottom, horizontal register. As the amount of charge in each photosite is extremely weak, it is necessary to amplify the signal. In a CCD array, signals are amplified after the horizontal register. However, in a CMOS array, amplifiers are embedded in each photosite using the same technology (photolithography) used to make integrated circuits. CMOS arrays are more sensitive and generate clearer images than CCD arrays.

To generate RGB color images, a *Bayer filter* is typically used, where each photosite (pixel) can receive only a single color (R, G, or B). The ratio of each color element is set to 1 red, 2 green, and 1 blue (corresponding to a 2×2 unit array), as the human eye is more sensitive to green color than to red and blue color. More details on smartphone camera are described in Chapter 9, Section 2.

3. White LED flash

In addition, most digital cameras are equipped with *flash*, enabling the ability to capture images in dim or dark environments. Although many cellular phones and very early smartphones were not equipped with flash, most recent smartphones have one. In the era of analog cameras, light bulbs were used as flash. These days, *white light emitting diodes (LEDs)*, are almost exclusively used as flashes for smartphones.

Light spectra of various light sources are shown in Fig. 2.4 [1,2]. The light spectrum of sunlight covers the whole range of visible wavelengths, which is a mixture of all colors. Such light is called as *polychromatic*. When all colors are represented, it looks white. As flash is used to replace natural sunlight, the light spectrum of a flash

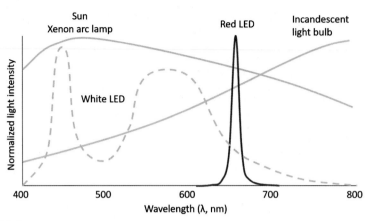

FIGURE 2.4

Light spectra of various light sources: sun and xenon arc lamp (solid gray), white LED (dashed gray), red LED (red), and incandescent light bulb (yellow).

should be similar to that of sunlight. Back in the old days, *incandescent light bulbs* were used the most for indoor lighting applications. The light spectrum of incandescent light bulbs as shown in Fig. 2.4, however, are somewhat different from that of sunlight, in that it is more shifted toward red color, resulting in a yellow coloration. Therefore, incandescent light bulbs are not an optimum choice for a flash in cameras. *Xenon arc lamps* have been popularly used as flashes, typically dubbed as *flashtube* or *flashlamp*, as its spectrum is quite similar to that of sunlight.

However, Xenon arc lamps are substantially bulky in size, typically about one-half or one-third the size of a standalone camera and require a separate battery to act as a power supply. As such, there is no way to incorporate this Xenon arc lamp into a cellular phone or smartphone. A better alternative is the use of *LED*. Both LEDs and photodiodes are made by joining p-type and n-type semiconductors together. Instead of applying a reverse-bias voltage to a diode to detect light, that is, photoconductive mode of photodiode, a forward-bias voltage is applied. As extra holes and free electrons flow in to the LED, they merge, disappear, and generate photons, that is, emit light. Different from the bulbs and lamps described earlier, LEDs generates relatively *monochromatic* light. The light spectrum of LEDs, as shown in Fig. 2.4, show only a single, relatively narrow peak. The peak wavelength (and subsequently its color) is determined by the type of semiconductor materials used in constructing an LED. LEDs are very small, consume a very small amount of electric current, and subsequently produce substantially smaller amounts of heat, as compared to incandescent light bulbs and Xenon arc lamp. However, LEDs cannot be used for flash or general indoor lighting applications, as they are monochromatic, that is, generate only a single color.

Recently, white LEDs have emerged as an alternative [2]. Conceptually, white LED is a hybrid of two different LEDs, emitting two different colors of blue color and yellow color (at the boundary of green and red). Thus, it covers all three RGB colors in the visible spectrum. Although the resulting spectrum of a white LED is not as flat as sunlight or Xenon arc lamp, the human eye cannot tell the difference as all three basic colors are represented. At this time of writing, all smartphone models utilize white LEDs as their flash. In fact, the same white LEDs are also popularly used for indoor lighting, called as *LED bulb* or *LED lamp*.

4. Photometry versus spectrometry

In optical biosensing, detection is made by either photometry or spectrometry. In *photometry*, light intensity is measured at a specific wavelength. Such light intensity can then be related to the concentration of the target in a sample. In *spectrometry*, light intensity is measured across a range of wavelengths, and the intensity−wavelength plot is obtained. The resulting plot is called as a spectrum, as shown in the previous sections. The location of the peaks can be related to the presence of certain chemical groups. Therefore, photometry is primarily used for quantification (how many of them?) while spectrometry is used for qualification (what is it?). In an analytical instrumentation laboratory, a single instrument is typically available that can perform both photometry and spectrometry, called as *spectrophotometer*. The sizes of typical spectrophotometers are substantial (typically bigger than most laptop computers), relatively expensive (over US $5000), and require professional training.

Spectrophotometers consist of (1) light source, (2) monochromator, (3) sample container (cuvette), and (4) photodetector, as shown in Fig. 2.5 in the top [1]. A *monochromator* is the device that delivers monochromatic light (at a specific wavelength) from a "white" light source. Through mechanical systems, a monochromator can continuously change the wavelength of incident light, and a light intensity + wavelength data pair is obtained one by one. In the past, such "scanning" took a minute or two to obtain a spectrum. Recently, scanning time has sufficiently been reduced well under 30 s, although it is still not an instantaneous process.

A better alternative is to place the monochromator (typically using *diffraction grating*) after the sample, as shown in Fig. 2.5 in the bottom. "White" light directly hits the sample, and the sample transmits the white light where certain colors are attenuated. Diffraction grating spreads such mixed light instantaneously, like a prism. The simple photodetector should also be replaced with a 1D CCD or CMOS array [1], where each pixel corresponds to a specific wavelength. Different from the conventional spectrophotometer, a full spectrum can be obtained instantaneously. As there is no moving component, it is possible to make the spectrophotometer substantially smaller, usually handheld. A *miniature spectrophotometer* from Ocean Optics is a good example [3].

Smartphones are equipped with a light source (white LED flash) and a photodetector (CMOS array). It may be possible to attach a monochromator (diffraction grating) to the smartphone's camera to make the setup similar to the one

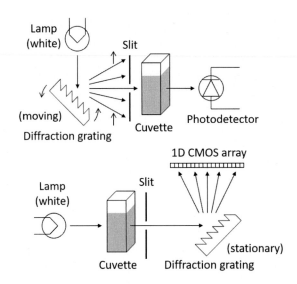

FIGURE 2.5

Two different setups of spectrophotometer. Top: white light—monochromator—cuvette—photodetector. Bottom: white light—cuvette—diffraction grating—photodetector. Collimating lens and mirror(s) are typically added to this system.

shown in Fig. 2.5 on the bottom. Although the smartphone's camera offers a 2D image, only 1D data will be utilized to construct a spectrum. However, such detection has been demonstrated relatively rarely. In addition to the requirement of a monochromator (diffraction grating), all components must be aligned with high precision, as a dislodgement of the diffraction grating in a sub-mm scale results in the shift in several hundreds of nm in the spectrum. Because of this, photometry has been popularly demonstrated in smartphone-based optical sensing but not spectrometry.

In smartphone-based photometry, a monochromatic light source is necessary, preferably without using monochromator (diffraction grating). Unfortunately, a smartphone's light source—white LED flash—is incapable of generating monochromatic light. However, a smartphone camera is capable of distinguishing different colors (ranges of wavelength, that is, red, green, and blue), although incapable of distinguishing specific wavelengths. As a result, at a cost of compromised accuracy, photometry becomes possible with a smartphone without using a monochromator (diffraction grating), optical filter, or separate light source.

If the photometry needs to be performed at a precise wavelength, a separate, monochromatic light source can be used, for example, a *laser diode*. The peak width of laser diode's spectrum is narrower than that of an LED, serving as an excellent monochromatic light source. Alternatively, an *optical filter* can be taped or attached to the white LED flash of the smartphone, allowing to generate monochromatic light.

Such optical filters can also be taped to the camera of the smartphone to detect only a specific wavelength, which is very useful in fluorescence detection.

5. Absorbance and colorimetry

Fig. 2.4 shows one type of spectra, which is a plot of the light intensities of light sources against wavelength. In practical photometry, the plots of the light intensities transmitted from samples against wavelength are more important. Samples are typically aqueous solutions in a transparent rectangular container ("*cuvette*"), but they can also be aqueous solutions in a microfluidic channel (in a *microfluidic device* or *lab on a chip*), in a *paper strip*, or even the solid surface of a biological sample. If the sample platform is not transparent, the light intensities reflected from samples can also be used. In both cases—transmitted and reflected—the light intensities are attenuated at certain ranges of wavelengths while they remain the same at other wavelength ranges. This is due to the selective *absorption* of certain ranges of wavelength (and subsequently color) by molecules in the sample. Such selective absorption leads to the development of certain coloration of a sample, under white light source.

Absorbance (A) is defined as the ratio of incident light intensity (I_0) and the transmitted (sometimes reflected) light intensity (I):

$$A = -\log I/I_0 = \log I_0/I$$

I_0 can be measured simply by removing the sample (shown in Fig. 2.5). Absorbance is correlated to the concentration of a target molecule in a sample using the well-known Beer—Lambert law:

$$A = \varepsilon \cdot b \cdot c$$

where ε is the *molar absorptivity*, b is the path length (width of a cuvette, a microchannel, etc.), and c is the molar concentration. Mass concentration can also be used, where ε should be switched to the mass absorptivity. Although it is possible to define ε for each target molecule, there are many other factors affecting the absorbance readings, for example, absorption by the air, the sample container (cuvette, microchannel, paper strip, etc.), nontarget molecules, and water. Therefore, a standard curve is normally constructed for a series of target solutions varying its concentration. Here, the target concentration in an unknown sample is back-calculated using the standard curve equation. This procedure is well documented in general chemistry and analytical chemistry textbooks [4].

Fig. 2.6 shows several examples of absorption spectra. The first example is of *gold nanoparticles* (quite popularly used as a dye/indicator for various optical biosensors, including pregnancy tests), showing very high green absorption, medium blue absorption, and almost zero red absorption [1,5]. The resulting coloration should be nonattenuated red (strong red) + medium attenuated blue (medium blue) + mostly attenuated green (no green), that is, pink coloration. Considering

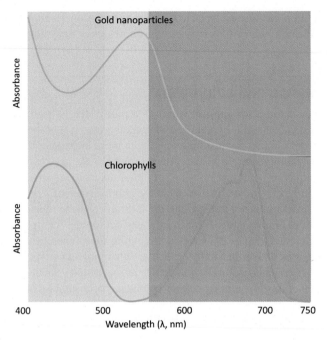

FIGURE 2.6

Absorption spectra of gold nanoparticles (top) and chlorophylls (bottom). The bottom spectrum includes the absorption by both types of chlorophylls (A and B).

the Beer–Lambert law, it would be wise to conduct the photometry at 530 nm (green) to find out the concentration of gold nanoparticles. It can also be done at blue color, although at a cost of reduced sensitivity.

The second example is *chlorophyll* (mostly found in plant leaves), showing very strong blue absorption, zero green absorption, and very strong red absorption (although only for very long wavelengths) [6]. The resulting coloration should be nonattenuated green (strong green) + medium attenuated red (medium red) + mostly attenuated blue (no blue), that is, green to yellow coloration. In addition, photometry should be performed at either blue colors or long-wavelength red colors to determine chlorophyll concentration.

As explained in the previous section, it is not possible to conduct photometry using smartphone at a specific, narrow-band wavelength, unless a separate monochromator or a separate monochromatic light source is used. In most cases, however, smartphones have shown satisfactory sensitivity and accuracy in identifying and quantifying various target molecules in a wide variety of platforms (cuvettes, paper strips, microfluidic channels, etc.), as they can determine the ranges of wavelength, that is, red, green, and blue color. Such a simplified analysis is called as *colorimetry*, and has most popularly been demonstrated with smartphone-based optical sensing.

Reflectance is also popularly used in lieu of absorbance when the platform is optically opaque, as is the case with paper strips (and paper microfluidic devices). Reflection spectrum is very similar to absorption spectrum, and its analytical procedure is quite similar to that of absorption spectrum. Identical colorimetric analysis can be performed with reflectance measurements.

Colorimetric detection from cuvettes, *ELISA* (*enzyme-linked immunosorbent assay*) plates, paper strips, microfluidic devices, etc. have all been well studied and demonstrated using smartphones [5,7,8]. In Chapter 4, Section 3, smartphone-based colorimetric sensing of glucose from whole blood or body fluids is demonstrated on paper strips and paper microfluidic devices (also known as *microfluidic paper analytic devices* or *μPADs*). In Chapter 6, smartphone-based colorimetric sensing of various target molecules is demonstrated on rapid kits. In Chapter 7, Section 2.1, smartphone-based colorimetric sensing of various biomarkers is demonstrated on both silicone-based and paper-based microfluidic devices. In Chapter 11, Sections 4.1 and 4.2, smartphone-based colorimetric sensing of allergens and antibiotics from food is demonstrated on cuvettes, ELISA plates, LFAs, and microfluidic devices.

Another example of colorimetric detection is *pulse oximetry.* A typical pulse oximeter looks like a clip (Fig. 2.7 inset), with red and *near infrared* (*NIR*) LEDs embedded inside the meter. It is able to record (1) *heart rate* or *pulse*, and (2) *oxygen saturation in blood* (*SpO$_2$*), hence the name *pulse oximeter* [1]. The red coloration of blood comes mostly from *red blood cells* (*RBCs*), which are essentially packs of *hemoglobin,* which is the protein responsible for delivering oxygen. The absorption spectrum of hemoglobin is altered depending on whether they carry oxygen (*oxyhemoglobin*) or not (*deoxyhemoglobin*).

The ratio of oxy- and deoxyhemoglobin can be obtained by measuring absorbance at 660 nm, thus photometric detection. However, the amount of blood

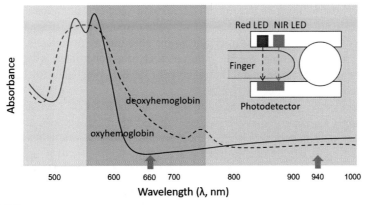

FIGURE 2.7

Absorption spectra of oxy- and deoxyhemoglobin, shown together with the schematic diagram of a pulse oximeter.

between the two sides of a clip can be varied by the thickness of the finger and from person to person. Therefore, a reference absorbance peak that is not altered by the blood's oxygenated status is needed. As shown in Fig. 2.7, the absorbance at 940 nm (NIR) can be used as such a reference. The resulting ratio of absorbance at 660 nm and 940 nm can be correlated to the ratio of oxy- and deoxyhemoglobin, and thus oxygen saturation rate ($\%$ SpO_2). In addition, the absorbance signals from a finger are pulsating due to the heartbeat, so such pulsation can be used to count the heart rate or pulse.

In the earlier era of smartphones, there was a software application used to measure the pulse through placing a finger on top of the smartphone's camera, although it was incapable of measuring SpO_2 due to the lack of the light source on the other side of finger and the lack of an NIR light source. Activity trackers and smartwatches can measure pulse in much more accurate way, as they are equipped with NIR light source and sensors.

Both SpO_2 and pulse can be monitored using a sophisticated sensor system, most notably *wearable sensors*. The sensor readings are normally transmitted wirelessly to smartphones. Details and examples are well documented and summarized in Chapter 10.

6. Fluorescence

In colorimetry, absorption for a range of wavelengths (i.e., color) is measured to relate it to the concentration of a target molecule. The target molecule itself can exhibit a specific coloration that is different from the other molecules in a sample. Chlorophyll and hemoglobin, described in the previous section, are good examples. Although such distinction may be possible if the absorption is measured at a specific wavelength, it is quite difficult to do so with only three choices of color (RGB). Moreover, many target molecules do not exhibit specific coloration—they either absorb all colors (opaque) or transmit all colors (transparent). In such case, it is necessary to "label" the target molecule with a colorimetric dye, that is, a molecule that exhibits a very strong, specific coloration. Gold nanoparticles, which are popularly used in rapid kits or lateral flow assays (LFAs; e.g., pregnancy test), are a good example of such colorimetric dyes.

In fact, there are better labeling dyes that have been popularly used in biosensing—*fluorescent dyes*. These are organic chemical compounds, normally containing multiple aromatic rings in their structure. They absorb light at a very specific wavelength, similar to colorimetric dyes. For colorimetric dyes, such absorbed light is converted to other forms of energy, most notably heat. Fluorescent dyes, however, release such absorbed light back, although the wavelength of such emitted light is slightly longer than the absorbed light. As the energy of light $E = hc/\lambda$, longer wavelengths represent lower energy. Such loss of energy is due to the phenomenon known as *Stokes shift* [1,2,4].

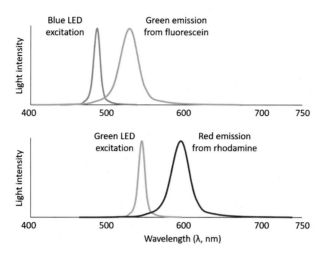

FIGURE 2.8

Fluorescence excitation—emission spectra of fluorescein (top) and rhodamine (bottom).

Fig. 2.8 shows examples of two popular fluorescent dyes. Fluorescein absorbs blue color (*excitation*) and releases green color (*emission*), while Rhodamine is excited with green color and emits red color. In colorimetric detection, the absorption spectra of target and nontarget molecules often overlap each other, making quantification difficult. In fluorescent detection, the fluorescent dye can selectively be excited, and the resulting fluorescent emission can be isolated easily. For the case of fluorescein, a narrow-band blue color is irradiated to the sample. Although there may exist some other molecules in the sample that may adsorb the same color, they cannot be detected as they do not emit green color.

As shown in Fig. 2.8, the excitation and emission peaks are overlapping each other to a smaller extent. Therefore, it becomes necessary to use *optical filters*, although this is not necessary, especially when a spectrometer is used. For the case of fluorescein, a *low-pass* optical filter at 500 nm can be used on the blue LED excitation, to eliminate any wavelength longer than 500 nm. Similarly, a *high-pass* optical filter at 500 nm can be used on the green emission detection, to eliminate any wavelength shorter than 500 nm [8]. In practical uses, a *band pass* (*BP*) filter is commonly used, where a low-pass and high-pass filter are sandwiched together.

To use a smartphone for fluorescence detection, it is almost mandatory to use optical filters, as a clear distinction between the excitation and emission peaks is difficult. Smartphones are only capable of performing colorimetric assays, not spectrometric ones. Fig. 2.9 shows several different examples of utilizing smartphones for fluorescence biosensing [9]. A single smartphone can be used as a fluorescent excitation light source (utilizing its white LED flash) and a fluorescent emission detector (utilizing its camera), if two different BP filters are attached to the flash and camera. Fluorescence can be detected from either a microfluidic device

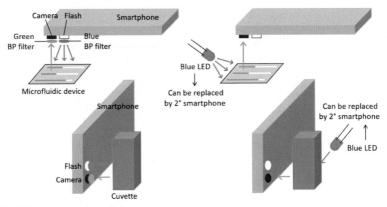

FIGURE 2.9

Smartphone detection of fluorescence from a microfluidic device or a cuvette.

(left top) or a cuvette (left bottom) in a reflective manner. This "reflective" setup is the only possible layout, as the flash and camera are on the same side of a smartphone. A separate light source can be used, typically located perpendicular to the light detector, to avoid possible reflection at the surface of the microfluidic device or cuvette, as shown in Fig. 2.9 to the right. A single LED is typically used for such separate light source, where a separate battery and a resistor (to protect the LED from being damaged by high current) are required. Alternatively, a secondary smartphone can be used solely as a light source, utilizing its flash.

Smartphone-based fluorescence detection from microfluidic devices has been widely investigated and summarized in Chapter 7, Section 2. Detection of bacteria, viruses, and pesticides from food samples has also been well documented using fluorescent dye-labeled antibodies, which is summarized in Chapter 11, Section 4.

7. Scattering

In previous sections, we learned that incident light can be either absorbed or transmitted. We also learned that fluorescence is a combination of both, where absorption and emission occur at different wavelengths. This statement is not exactly true, as the incident light can also be *scattered*. Similar to fluorescence emission, scattering occurs in every possible direction, which can be detected typically perpendicular to the incident light. Therefore, the intensity of the incident light should be equal to the sum of the absorbed, transmitted, and scattered light ($I_{incident} = I_{absorbed} + I_{transmitted} + I_{scattered}$).

As the wavelengths of incident and scattered light are identical to each other, scattering is in general not very useful compared to fluorescence. When the scattering object becomes sufficiently large in size (comparable to the wavelength of light), an interesting phenomenon can be observed. At certain scattering angles,

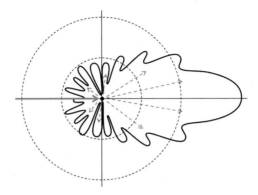

FIGURE 2.10

A plot of Mie scattering intensity against the scattering angle in a polar coordinate system. A scattering object (particle) is located at the center. Incident light (solid arrow) is delivered from the left to the particle in the center, which is scattered to many directions (dotted arrows).

the scattering is constructively added together, while at other angles the scattering is destructively attenuated. As a result, the scattering intensity "oscillates" over the scattering angle, and the location of such oscillation is varied by the size and refractive index of the scattering object. This is called as *Mie scattering* [10]. The plot of scattering intensity over scattering angle is shown in Fig. 2.10.

Mie scattering can be utilized for particle immunoassays, in which submicron particles (diameter of 0.1–1 μm; thus, their diameter is comparable to the UV, visible, and NIR wavelengths) are conjugated with antibodies to target molecules (proteins, viruses, bacteria, etc.) This particle suspension is mixed with a sample, and the presence of target molecules should induce the particles to aggregate via antibody–antigen binding. This particular phenomenon is called as *particle immunoagglutination* (Fig. 2.11). As the effective particle diameter changes upon such immunoagglutination, the Mie scattering pattern should also be altered. When the Mie scattering is measured at a fixed scattering angle, this can be related to the increase or decrease in scattering intensity, and this change can be used to calculate the target concentration in the sample. Mie scattering detection from microfluidic devices has been extensively demonstrated using smartphones, as shown in Fig. 2.10 [11–13]. The angle of incident light and scattering detection is optimized for maximum scattering from particles and minimum scattering from other molecules, paper fibers, as well as direct reflection.

Smartphone-based Mie scatter detection is further explained in Chapter 11, Section 4, where *E. coli* and *Salmonella* were detected for food safety applications.

In addition, scattering is a very popular method for counting cells in a capillary tube (*flow cytometry*), along with fluorescence. Such scattering (and fluorescence) detection from a flow cytometer has also been well studied using a smartphone camera (and flash), which is summarized in Chapter 5.

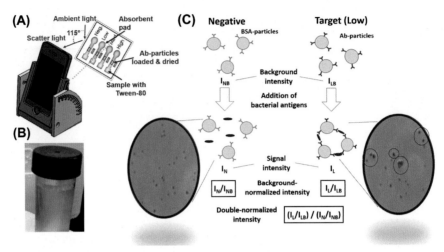

FIGURE 2.11

Mie scattering detection of particle immunoagglutination assay on paper microfluidic device (μPAD). Smartphone is positioned to detect optimum Mie scattering from a μPAD. *E. coli* was detected from human urine sample.

Reprinted from Ref. [13] with permission. (C) 2015 Elsevier BV.

8. Microscope attachment

Before the era of smartphones, many digital cameras were capable of digital zoom and optical zoom. Although digital zoom makes the image larger at the cost of sacrificing its resolution, optical zoom makes the image larger without sacrificing the resolution through adjusting the distance between two objective lenses. In actuality, such optical zoom is the working principle of many different optical microscopes. In the past, the optical zoom in many digital cameras was limited to $2\times-10\times$. Nowadays, standalone digital cameras (e.g., DSLR cameras) provide extremely high resolutions (16–18 MP) and enhanced optical zoom features ($30\times-50\times$), primarily to outperform smartphone cameras.

Although optical zoom has traditionally not been available in smartphone cameras, it has recently become available, although very limited, for example, $2\times$ optical zoom in the most recent iPhone and Galaxy phones. As the magnification in typical laboratory microscopes is much higher than the optical zoom available in these cameras, it becomes necessary to use an attachment device to convert a smartphone into a microscope. Such attachments are typically called as *microscope attachments* to a smartphone.

A wide variety of microscope attachments are commercially available. Most of them look quite similar to the objective lenses used for a laboratory microscope. Such attachments can be either clipped or slid on to a smartphone, positioned on top of its camera. Although some attachments are designed to deliver

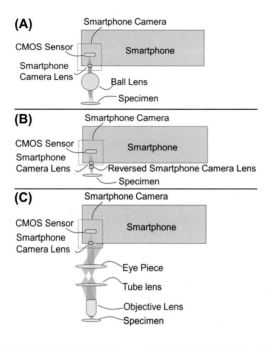

FIGURE 2.12

Schematic illustration of three different microscope attachments to smartphone. Identical to Fig. 2.7 in Chapter 9.

the flash light from a smartphone to the sample, most of them are equipped with a separate white LED light source that can irradiate right on top of the sample. Such white LEDs are typically powered by a small battery incorporated within the attachment (Fig. 2.12).

More sophisticated microscope attachments are under development. Very high magnifications, fluorescence imaging, and even confocal microscopy have been demonstrated. Such recent accomplishments are summarized in Chapter 9.

9. Illumination/ambient light sensor

Besides front and rear cameras, smartphones are also equipped with and additional light sensor—called as *illumination sensor* or *ambient light sensor*. It senses the level of overall ambient lighting so that the brightness of the phone's screen can automatically be adjusted. Although it is incapable of recognizing color, it is simple enough to convert the signal directly into sensor readings. With a camera, on the other hand, a separate image processing code must be developed to evaluate the light intensity from the area of interest in the acquired image. Due to this simplicity, several attempts have been made to utilize this illumination/ambient lighting sensor as an optical detector. Such attempts are summarized in Chapter 7, Section 3.

Illumination/ambient light sensors can also be used in conjunction with the camera, to compensate for the errors caused by ambient lighting [14].

References

[1]   Yoon J-Y. Introduction to biosensors: from electric circuits to immunosensors. 2nd ed. New York, NY: Springer; 2016.

[2]   McCracken KE, Yoon J-Y. Recent approaches for optical smartphone sensing in resource-limited settings: a brief review. Anal Meth 2016;8:6591−601.

[3]   Ocean Optics. Flame miniature spectrometer user manual. Dunedin, FL: Ocean Optics; 2016 [Chapter 5]. Available at: https://oceanoptics.com/wp-content/uploads/FlameIO.pdf.

[4]   Christian GD, Dasgupta PK, Schug KA. Analytical chemistry. 7th ed. Hoboken, NJ: John Wiley & Sons; 2013.

[5]   You DJ, Park TS, Yoon J-Y. Cell-phone-based measurement of TSH using Mie scatter optimized lateral flow assays. Biosens Bioelectron 2013;40:180−5.

[6]   Chung S, Breshears LE, Yoon J-Y. Smartphone near infrared monitoring of plant stress. Comput Electron Agric 2018;154:93−8.

[7]   Kaarj K, Akarapipad P, Yoon J-Y. Simpler, faster, and sensitive Zika virus assay using smartphone detection of loop-mediated isothermal amplification on paper microfluidic chips. Sci Rep 2018;8:12438.

[8]   Ulep T-H, Yoon J-Y. Challenges in paper-based fluorogenic optical sensing with smartphones. Nano Converg 2018;5:14.

[9]   McCracken KE, Tat T, Paz V, Yoon J-Y. Smartphone-based fluorescence detection of bisphenol a from water samples. RSC Adv 2017;7:9237−43.

[10] Nicolini AM, Toth TD, Kim SY, Mandel MA, Galbraith DW, Yoon J-Y. Mie scatter and interfacial tension based real-time quantification of colloidal emulsion nucleic acid amplification. Adv Biosyst 2017;1:1700098.

[11] Park TS, Li W, McCracken KE, Yoon J-Y. Smartphone quantifies Salmonella from paper microfluidics. Lab Chip 2013;13:4832−40.

[12] Park TS, Yoon J-Y. Smartphone detection of *Escherichia coli* from field water samples on paper microfluidics. IEEE Sens J 2015;15:1902−7.

[13] Cho S, Park TS, Nahapetian TG, Yoon J-Y. Smartphone-based, sensitive μPAD detection of urinary tract infection and gonorrhea. Biosens Bioelectron 2015;74:601−11.

[14] McCracken KE, Angus SV, Reynolds KA, Yoon J-Y. Multimodal imaging and lighting bias correction for improved μPAD-based water quality monitoring via smartphones. Sci Rep 2016;6:27529.

Basic principles of electrochemical biosensing using a smartphone

3

Jeong-Yeol Yoon, PhD

Professor, Department of Biomedical Engineering and Department of Biosystems Engineering, The University of Arizona, Tucson, AZ, United States

In Chapter 2, one of two most popularly used transducers for biosensors—optical—was explained, and how such optical biosensing can be demonstrated using a smartphone. In this chapter, the other transducer—electrochemical—will be covered and how such electrochemical biosensing can be demonstrated using a smartphone.

1. Ohm's law

Electrochemical biosensors are classified into three categories, depending on what is measured as their signals. They are (1) potentiometric, (2) amperometric, and (3) conductometric biosensors, which measure voltage, current, and conductance (inverse of resistance), respectively [1]. Therefore, these three categories are closely related to the foundational basics of electrics and electronics: *Ohm's law*. Ohm's law is defined as follows:

$$V = IR$$

where V is electric voltage (in *volts*, V), I is electric current (in *amperes*, A), and R is electric resistance (in *Ohms*, Ω).

Fig. 3.1 shows a simple *electric circuit*, with a voltage source and two resistors connected in series. Typically, one of such resistors is a biosensor (the resistor in an ellipse). For the sake of simplicity, let us limit our discussion to the *direct current* (*DC*) voltage source (e.g., a battery or an AC-to-DC power adaptor). In this circuit, the voltage source provides a DC voltage of +3 V, which makes the current to flow through two resistors. Bear in mind that the voltage is a relative term, that is, the difference in electric potential between the power source's cathode (+) and anode (−). Although we can define the potential at cathode as 0 V and at anode as −3 V, creating the net potential difference (voltage) of +3 V, the more commonly used convention, is to make the voltage at the voltage source's anode as 0 V, that is, a reference point or *ground*, and that at the cathode as +3 V. With this definition,

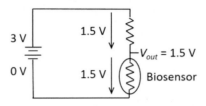

FIGURE 3.1

An electric circuit with a biosensor.

the voltage at the voltage source's cathode, +3 V, must be dropped down as the current passes through two resistors, and eventually to 0 V as it returns back to the voltage source's anode (ground). If the resistance of two resistors are equivalent, a half of 3 V (= 1.5 V) should be dropped after the first resistor, and the other half after the second resistor. The voltage at the point in between two resistors (with a reference to the power source's anode or the ground) should be 1.5 V (= V_{out}). If the first resistor is a fixed one and the second resistor is a biosensor, with resistance that varies with the biological variable being tested, V_{out} should be altered. If such V_{out} is measured as a biosensor readout, such detection is referred to as potentiometric electrochemical biosensor or simply *potentiometric biosensor*. If the resistance of a biosensor is fixed while the current flowing through a biosensor is varied, it will be necessary to measure current flowing through the biosensor. Such detection is referred to as amperometric electrochemical biosensor or simply *amperometric biosensor*. It is also possible to directly measure the change in biosensor's resistance. The inverse of resistance is referred to as *conductance*, and such detection is referred to as conductometric electrochemical biosensor or simply *conductometric biosensor*.

2. Potentiometric biosensor: Nernst equation

The most studied and commonly used potentiometric sensor is *ion-selective electrode (ISE)*. Although it is strictly not a biosensor as it measures the concentration of a specific ion in an aqueous solution, ISE is commonly used in various biosensing applications. ISE is essentially a glass tube filled with a reference electrolyte solution, with an ion-selective membrane at its end, as shown in Fig. 3.2. The ion-selective membrane allows only the target ions to move in and out, while preventing other ions from doing so. Such small change in the target ion concentration results in the change in electric voltage. The voltage can be related to the target ion's concentration, using the *Nernst equation* [1–3]:

$$\Delta E = (RT/nF)\ln\left[M^{+}\right]$$

where ΔE is the voltage difference of ISE (in V), R is gas constant (= 8.314 J/mol·K), n is the number of electrons (e.g., $n = 2$ for Ca^{2+}), F is Faraday constant

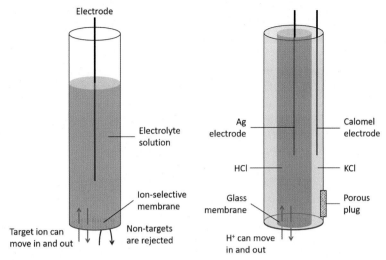

FIGURE 3.2

Schematic diagrams of an ion-selective electrode (ISE; left), showing a half-cell without a reference electrode, and a combined pH electrode, that is, with a reference calomel electrode (right).

(= 96,487 C/mol), and $[M^+]$ is the molar concentration of a target ion (in mol/ L = M). As ΔE is a function of a natural log (ln) of target ion's concentration, it covers quite a broad range of target concentration. For example, the dynamic range of potassium (K^+) ISE is from 1 μM to 1 M (equivalent to 0.04 ppm to 39,000 ppm) [4].

As ISEs can measure only the difference in electric potential (voltage), a reference electrode is necessary to correctly estimate the concentration of target ion. ISEs are essentially half-cells. Manufacturers occasionally offer *combination ISE*, where the half-cell ISE and the reference electrode are combined into a single device.

The following ions can be identified and quantified using ISEs: Ag^+, Cl^-, Br^-. SCN^-, S^{2-}, NO_3^-, Cu^{2+}, BF_4^-, ClO_4^-, K^+, and H^+ (this list is not exhaustive). The last one is a special case, since the molar concentration of H^+ is directly related to the pH:

$$pH = -\log\left[H^+\right]$$

The Nernst equation can be rewritten with base-10 log:

$$\Delta E = 2.303\,(RT/nF)\log\left[M^+\right]$$

Substituting the definition of pH, universal constants of R and F, room temperature $T = 298$ K (25°C), and $n = 1$ for H^+ into this equation gives [1−3].

$$\Delta E = -0.059\,pH$$

The ISE for quantifying [H^+] is specifically referred to as a *pH electrode*. Similar to combination ISE, a combination pH electrode (or *combined pH electrode*) is commercially available. In fact, it is the most common form of a pH electrode.

Both ISEs and pH electrodes provide voltage as their output. As indicated by the previous equation, one unit increase in pH results in a 0.059 V (=59 mV) decrease in voltage. However, as the current flowing through typical ISEs and pH electrodes tends be very low, voltage measurements must be made with caution. If a circuit shown in Fig. 3.1 is used, where the bottom biosensor (the resistor in an ellipse) is replaced with a combination ISE or a combined pH electrode, a current-limiting circuit will be necessary to minimize the current flowing through it. More common approach, however, is the use of an *operational amplifier (op-amp)* that can theoretically make the current flowing into the op-amp close to zero. A special type of op-amp is typically used for ISE and pH electrode, for example, *junction gate field-effect transistor*-based op-amp, which can make the input current extremely small [1].

3. Amperometric biosensor: Redox reaction

As explained in Chapter 1, enzyme can be used as a bioreceptor to identify and quantify the concentration of a target substrate. Enzyme is one type of protein, which functions as a biological catalyst, allowing chemical or biochemical reactions to occur at body temperature. Each enzyme binds to a specific target molecule, referred to as substrate. There are several classes of enzymes one of the most important being oxidoreductases. Once oxidoreductase enzyme specifically binds to a target substrate, it oxidizes the substrate while the enzyme itself is reduced to an inactive form. Substrates are usually small biomolecules, for example, glucose, cholesterol, alcohol, lactic acid, uric acid, urea, etc.

It is possible to make this reduction—oxidation (*redox*) reaction into a cycle, if an *electronic mediator* is used. One of the first such attempts was made in detecting glucose (target substrate), using the enzyme *glucose oxidase (GOx)* and *ferricyanide ion* $Fe(CN)_6^{3-}$ as an electronic mediator. First, glucose oxidase catalyzes the oxidation of glucose in the presence of dissolved oxygen in the solution (e.g., blood).

Glucose + O_2 + GOx (oxidized) — > Gluconic acid (oxidized)
+ GOx (reduced) + H_2O_2 (byproduct)

The electron mediator ferricyanide can convert the reduced GOx (inactive form) back to the oxidized GOx (active form), while ferricyanide is reduced to *ferrocyanide* $Fe(CN)_6^{4-}$.

GOx (reduced) + $Fe(CN)_6^{3-}$ (oxidized) — > GOx (oxidized)
+ $Fe(CN)_6^{4-}$ (reduced)

If electrodes are connected to this solution, the extra electron from ferrocyanide can be given back to the electrode, generating electric current, while ferrocyanide is oxidized back to ferricyanide (active form).

$$Fe(CN_6)^{4-} \text{ (reduced)} \longrightarrow Fe(CN_6)^{3-} \text{ (oxidized)} + e^-$$

This cycle can be repeated, generating detectable electric current. As this current is a function of the concentration of glucose in the sample (blood), it can be used to estimate the blood glucose concentration. In practical applications, more sensitive detection can be made if a voltage is applied to the system. This situation is identical to the photoconductive mode of photodiode described in Chapter 2.

Many other electronic mediators have been investigated and evaluated, including flavin adenine dinucleotide (FAD^+), nicotinamide adenine dinucleotide (NAD^+), and *pyrroloquinoline quinone* (*PQQ*). The latter two are used with a different enzyme, *glucose dehydrogenase* (*GDH*). At the time of writing, the de facto standard for commercial glucose sensing is the use of GDH-PQQ, as GDH does not need dissolved oxygen from blood and the electron transfer rate of PQQ is very fast [1].

$$Glucose + GDH\text{-}PQQ \text{ (oxidized)} \longrightarrow Gluconolactone \text{ (oxidized)} \\ + GDH\text{-}PQQ \text{ (reduced)}$$

$$GDH\text{-}PQQ \text{ (reduced)} \longrightarrow GDH\text{-}PQQ \text{ (oxidized)} + H^+ + 2e^-$$

Amperometric biosensing has been demonstrated for a wide variety of target substrates, mostly small biomolecules: cholesterol using cholesterol oxidase (ChOx), ethanol using alcohol dehydrogenase, lactic acid using lactic oxidase, uric acid using uricase, urea using urease, etc.

Electrochemical sensing of glucose from whole blood is currently one of the biggest applications in whole biosensor market. Monitoring blood glucose concentration is critical for patients with diabetes, so that the patient can make an informed decision regarding meals, exercise, and diabetic pills, as well as injecting insulin if necessary. Fig. 3.3 shows a series of photographs demonstrating the use of a commercial glucose meter from a *finger prick* sample.

Further details on commercial glucose sensing (mostly electrochemical) are summarized in Chapter 4.

4. Conductometric biosensor

It is also possible to measure resistance or conductance (inverse of resistance) as a biosensor readout. Resistance or conductance is typically measured by dipping two metal plates into the sample solution (Fig. 3.4 in the left). As resistance or conductance can be altered by the size of the metal plate and the distance between two plates, it becomes necessary to normalize resistance or conductance. Such normalized resistance or conductance is referred to as resistivity or conductivity.

FIGURE 3.3

Use of a commercial glucose meter for monitoring blood glucose level. A lancet device is pricking the patient's finger, generating a small volume of blood on it. A glucose strip (where the enzyme and electronic mediator are preloaded) is inserted into a meter, following by contacting the end of a strip to the blood sample. Blood flows into the strip spontaneously via capillary action, and the redox reaction cycle occurs under an applied voltage. Change in electric current is recorded, converted to the blood glucose concentration, and displayed on the meter's screen.

Reprinted from Ref. [1] with permission, (C) 2016 Springer.

Resistivity is defined as RA/L, where R is the resistance in Ω, A is the area of metal plate in cm^2, and L is the distance between two metal plates in cm. The resulting unit of resistivity is $\Omega \cdot cm$. Similarly, conductivity is defined as L/RA, and the resulting unit is $\Omega^{-1} \cdot cm^{-1}$. The unit Ω^{-1} is specifically referred to as Siemens (S), hence the unit becomes S/cm [1–3].

As the changes in resistivity or conductivity in a solution during chemical and biochemical reactions are very small, it is difficult to measure such subtle changes in a conventional manner. A differential measurement is required, and *Wheatstone bridge* can make such differential measurement (Fig. 3.4 in the right) [1]. In a Wheatstone bridge, four resistors are connected in a diamond configuration. It can be considered as two branches in parallel, where each branch contains two resistors connected in series. If four resistors have identical resistance, with the input voltage V_{in} of 3 V, the voltages at the points a and b (V_a and V_b) should be 1.5 V (both with reference to the ground), as all resistors are identical. The voltage difference between points a and b ($= V_a - V_b = V_{out}$) should be 0 V. In this situation, the bridge

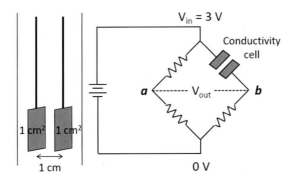

FIGURE 3.4

Left: A conductivity cell is essentially two metal plates that are immersed into a solution. This specific conductivity cell has two metal plates with area of 1 cm² each that are separated by 1 cm from each other. Right: In a Wheatstone bridge, four resistors are connected in a diamond configuration, where the top right resistor has unknown resistance and the other three's resistances are known. In this diagram, a conductivity cell is an unknown resistor.

is referred to as *balanced*. If the top right resistor's resistance is changed slightly and the other three resistors are unchanged, V_b will be changed slightly, for example 1.5001 V, while V_a is unchanged, 1.5 V. A typical multimeter cannot make a distinction between 1.5 V and 1.5001 V. $V_a - V_b = V_{out}$, on the other hand, can easily be measured by a multimeter if measured in mV (0.1 mV) or μV scale (100 μV). This V_{out}, in turn, can be used to back-calculate the resistance change of the top right resistor. Conductivity cell can be placed to the top right resistor and measure such subtle changes in resistivity or conductivity.

Conductivity cell itself is not very useful in biosensing. However, the surface of each electrode can be modified with bioreceptor, for example, antibodies. When a big target like bacterium or cell can bridge two electrodes together, conductivity will significantly increase. Although many different bioreceptor-modified conductometric biosensors have been developed, *interdigitated microelectrode* (*IME*) biosensor has received substantial interest from biosensor community [5]. In IME biosensor, two metal plates in a conductivity cell are modified into a pair of ladder-like electrode patterns printed on a surface (Fig. 3.5). Electrodes are preimmobilized with antibodies to the target, in this case, anti-*Escherichia coli*. The presence of *E. coli* in the sample results in binding of *E. coli* to the surface-bound antibodies, and eventually connecting two electrodes together. Conductivity will be changed, which can be measured by a Wheatstone bridge circuit.

As shown in Fig. 3.4, a DC voltage must be applied to the Wheatstone bridge. Biological samples, however, contain many biomolecules and enzymes, and the application of DC voltage may result in oxidation and/or reduction of molecules. Therefore, in most conductometric biosensors, *AC* (*alternating current*) voltage is preferred over DC. As the polarity of voltage is constantly changed in AC, the

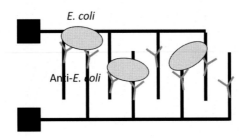

FIGURE 3.5

Interdigitated microelectrode (IME) biosensor, with surface-bound anti-*E. coli*, is detecting *E. coli*.

definition of resistance (and conductance) must be expanded using complex numbers. The AC analog of resistance is called impedance:

$$Z \text{ (impedance)} = R \text{ (resistance)} + jX \text{ (reactance)}$$

R is related the magnitude of oscillation and X to the phase shift of oscillation. j is the solution of the equation $x^2 = -1$ (sometimes represented as i). As there is no oscillation in DC, X becomes zero and impedance becomes the same as resistance with DC voltage. The units of impedance, resistance, and reactance are the same: Ohm (Ω). Similarly, the AC analog of conductance can be defined in a similar manner:

$$Y \text{ (admittance)} = G \text{ (conductance)} + jB \text{ (susceptance)}$$

$$Y = 1/Z$$

The units of admittance, conductance, and susceptance are the same: Siemens ($S = \Omega^{-1}$). Impedance and admittance measurements typically require a separate instrument, known as *impedance analyzer*, which has been bulky and expensive, although cheaper and portable impedance analyzers are emerging on the market in recent years.

5. Temperature sensors

Although temperature sensors are not really biosensors, measurement of body temperature is an essential practice in clinical diagnostics. Many biosensors are capable of detecting body temperature as their secondary sensor readings.

The oldest and still commonly used temperature sensor is a *thermocouple*. It is essentially two dissimilar metal wires joined together at the temperature sensing junction, as shown in Fig. 3.6 in the top [1]. If there is a difference in temperature between the junction and the other ends of metal wires, heat will flow through both wires, creating more free electrons. Although metals are generally considered conductors, they still have some electric resistance. The heat-induced electron flow

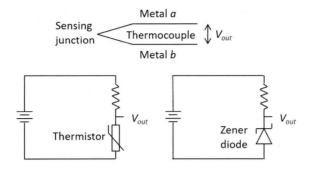

FIGURE 3.6

Temperature sensors: thermocouple (top), thermistor (bottom left), and zener diode (bottom right).

will certainly change the resistance of each metal wire. The difference between the voltage drops across the metal wire a (V_a) and the metal wire b (V_b) can be measured as its sensor reading (V_{out}), which is linearly proportional to the temperature at the sensing junction.

Although thermocouples are relatively accurate and cover a wide range of temperature (up to 1500°C), higher accuracy is necessary for biological and clinical applications, preferably less than 0.1°C resolution. To this end, semiconductor-based temperature sensors are preferred. All three types of semiconductor-based components can be used to sense temperature: resistor, diode, and transistor.

A resistor that is made sensitive to temperature is referred to as a *thermistor* (Fig. 3.6 in the bottom left) [1]. Similar to thermocouple, external temperature can alter the resistance of a thermistor. As the resistance value of a thermistor is much higher (i.e., less conductive) than that of the metal wires in a thermocouple (i.e., more conductive), thermistor is more sensitive to temperature than thermocouple. The downsides of thermistor are (1) narrower sensing range of temperature, which is not a big issue in measuring body temperature, and (2) nonlinear response, requiring an empirical fitting equation.

One type of diode is particularly useful and sensitive in measuring temperature: *zener diode*. Although zener diode was originally developed as a voltage regulator, it is also effective in measuring temperature [1]. In a reverse bias setup, as shown in Fig. 3.6 in the bottom right, zener diode does not allow electric current to flow, similar to other types of diodes. At a certain voltage, however, the current starts to flow with almost no resistance, referred to as zener voltage. (In fact, the same phenomenon occurs in regular diodes, while such voltage is very high and referred to as breakdown voltage, which may result in permanent damage.) The zener voltage is a function of temperature, which can be used for sensing temperature. The zener voltage is not a function of input current (within an acceptable range) and subsequently input voltage, which is a very useful feature when the power source does not generate constant voltage (e.g., battery). Similarly, transistor can also be used to sense temperature, although it is somewhat less common compared to zener diode.

These temperature sensors can be incorporated into a flexible substrate that can be attached to human skin. Such epidermal wireless thermal sensors (eWTS) are described in Chapter 10, Section 6. Additionally, IR cameras can also take thermal images of human skin, which can be used to estimate the local temperature of human body. Such IR cameras are particularly useful for monitoring wound, especially diabetic foot ulcer, described in Chapter 8, Section 3.4.

6. Analog-to-digital conversion and microcontrollers

Regardless of electrochemical detection methods, voltage, current, and conductance signals are eventually converted into voltage signals: Current signal can be converted into voltage through measuring a voltage drop across a known resistor ($I = V/R$) and conductance signal can also be converted into voltage utilizing a Wheatstone bridge. These voltage signals are analog signals, and must be converted into digital signals to be further processed. Such conversion is achieved by *analog-to-digital converter* or *A/D converter*. Back in the old days, A/D converters were typically a printed circuit board (PCB) that can be inserted into an expansion slot of a desktop computer. Digital data processing can be performed on a desktop computer. Both low-end and high-end A/D converters have been available, where the major differences between low and high ends were (1) resolution and dynamic range represented by 8-bit (0–255), 12-bit (0–4095), 16-bit (0–65,535), etc., (2) sampling rate represented by Hz or kHz, and (3) the number of channels, that is, how many data streams can be fed in simultaneously, for example, 2-, 4-, 8-, 16-channel, etc.

In commercial biosensors, the use of a separate desktop computer is discouraged. Therefore, an A/D converter, a data processing unit (typically a microprocessor), small memory, and a user interface (a small LCD screen) must be integrated into a single PCB and incorporated into a small enclosure. To this end, a separate integrated circuit (IC) dedicated for A/D conversion has been developed and widely used.

Later, several different microcontrollers have been suggested, developed, and manufactured, in an open-source and open-architecture manner. They are essentially very small computer in a single PCB, capable of receiving both analog and digital signals (in multiple channels) and transmitting the digitally processed results to other devices. One of the most popular microcontrollers is *Arduino microcontroller*, shown in Fig. 3.7. It is equipped with multiple pins capable of receiving digital signals as well as analog signals. For processing analog signals, an A/D converting IC is integrated on its board. Arduino has its own microprocessor and a small memory, and a compiled code can be uploaded to an Arduino via its *USB* (*universal serial bus*) port, to digitally process the input data. The processed results can be sent to other devices again through the USB port. Although Arduino has extensively been utilized and a wide variety of "accessories" are currently available, there still exist several limitations such as a limited memory space, insufficient computing power, etc. The real problem of Arduino is, in fact, the lack of user interface on its own.

FIGURE 3.7

Arduino microcontroller (left) and Raspberry Pi (right).

Reprinted from Refs. [6,7].

Typically, a small LCD screen can be connected to an Arduino to display the result. Codes are typically developed on a separate desktop or laptop computer, but not directly on an Arduino.

Raspberry Pi has emerged as a better alternative to Arduino, where it is equipped with faster microprocessor, more memory, multiple USB ports to connect a keyboard and/or a mouse, and most importantly an *HDMI* (*high-definition multimedia interface*) port to connect an LCD monitor (Fig. 3.7 in the right). Raspberry Pi is typically loaded with *Linux* operating system, and hence it can function as a standalone computer. Codes are typically developed directly on Raspberry Pi. Recently, microcontrollers can also be enhanced with WiFi and Bluetooth technology, that is, capable of transmitting the assay results in a wireless manner. As all smartphones are capable of WiFi and Bluetooth, a microcontroller can serve as an intermediary data acquisition/processing unit between an electrochemical biosensor and a smartphone.

Although microcontrollers are generally small, the overall dimension of smartphone-based electrochemical biosensor system can become substantial when all components are represented (electrochemical sensor + microcontroller + smartphone). Specifically when only a single data reading is necessary, both Arduino and Raspberry Pi microcontrollers can be considered overkill. In such a case, a single IC or a very small PCB can be utilized instead of a general-purpose microcontroller. Such "adaptors" are explained in Chapter 4, Section 4 for connecting amperometric glucose sensor to a smartphone and also in Chapter 7, Section 4 for connecting amperometric immunoassay microfluidic chip to a smartphone (Fig. 3.8) [8].

7. Micro-USB port

Once analog voltage signals are converted into digital signals using a circuit board "adaptor" (or a microcontroller), they can be transmitted directly to a smartphone. These days, all smartphones are equipped with a micro-USB port, which is a smaller-than-usual USB port to make the smartphone sufficiently thin (Fig. 3.8).

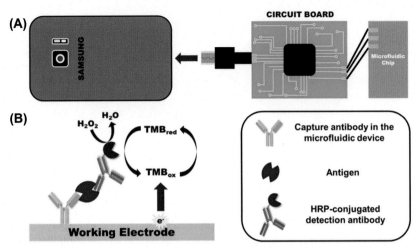

FIGURE 3.8

Amperometric electrochemical immunoassay microfluidic chip is connected to a circuit board, transmitting digitalized data to a smartphone via micro-USB port. Identical to Fig. 7.6 in Chapter 7.

Aside from its size, its functions are quite similar to regular USB ports. It can transmit digital data in very high speed, which is normally utilized to download photos, video clips, songs, etc. from a phone to a computer, or upload from a computer to a phone. It can also be used to charge the phone's battery with an AC-to-DC adaptor, as well as to draw battery power from a phone to an external device. The latter feature is important in working with "adaptors" and microcontrollers—they can simply draw power from a smartphone's battery. The processed digital signals can then be uploaded to a smartphone, and may further be processed within a smartphone app to display the assay results to the user. These results can also be uploaded to a cloud storage. As smartphone can offer further data processing, user interfacing, data sharing, and power supply, the adaptor can be made extremely simple with minimum size and cost.

8. Bluetooth and NFC

It is also possible for a circuit board "adaptor" to send the digital signals to a smartphone in a wireless manner. Most smartphones are capable of *Bluetooth*, which is essentially a short-distance radio wave data transmission. Many IC chips are available for Bluetooth capability, which can be incorporated into the above "adaptor." Fig. 3.9 shows an example of potentiometric electrochemical detection of kanamycin (antibiotic) using Bluetooth and a smartphone. More details are covered in Chapter 7, Section 4, Chapter 10, Sections 2, 3 and 5, and Chapter 11, Section 4.

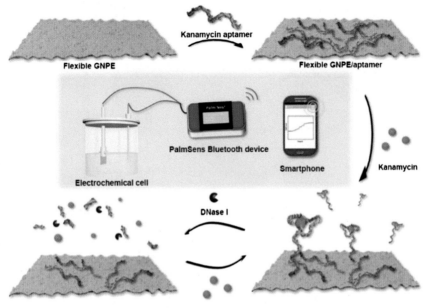

FIGURE 3.9

Potentiometric electrochemical aptamer assay of kanamycin (antibiotic) using Bluetooth and a smartphone. Identical to Fig. 11.10 in Chapter 11.

Reprinted from Ref. [9] with permission from Elsevier.

Another emerging form of wireless data transmission is near field communication (NFC). The major advantage of NFC over Bluetooth is that it can transfer both data and power, which is a significant benefit in most electrochemical biosensing adaptors, as described in the previous section (micro-USB port). However, NFC can be transmitted over only a very short distance (a couple of cm) while Bluetooth can be transmitted over about 10 m (class 2). In addition, not all smartphones are capable of NFC. At the time of writing, NFC and Bluetooth are complementary to each other, although it may be possible for NFC to dominate over Bluetooth in the future. NFC-based communication with smartphones is well summarized in Chapter 10.

In addition, pressure signals can also be transmitted via Bluetooth and potentially via NFC from a "wearable" pressure sensor. One example is described in Chapter 8, where the pressure from a wound (e.g., diabetic foot ulcer) is monitored from a pressure sensor mounted on socks or shoes.

9. Audio jack

All of the previously mentioned connections—micro-USB, Bluetooth, and NFC—requires an adaptor or a microcontroller to convert analog voltage signals into digital signals, as smartphone can recognize and process only digital data. However, there is

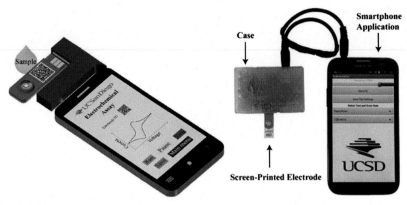

FIGURE 3.10

An electrochemical biosensor is connected to a smartphone via audio jack.

Reprinted from Refs. [10,11] with permission from IEEE and Elsevier.

one component in smartphone that deals with analog signals: *audio jack*. Audio jack is used to connect headphones (typically ear-fitting headphones or simply earphones), which are also equipped with microphone for voice calls. Microphones generate analog voltage signals in response to the sound wave, which are sent to the smartphone via an audio jack. Inside the smartphone, such analog signals are then converted to digital signals utilizing an on-board A/D converter. In the same manner, the audio signal from the other caller is also delivered to the headphones in an analog way.

It is possible to use this audio jack as an input terminal for transferring analog voltage signals to a smartphone. One such example is shown in Fig. 3.10 [10,11], where the voltage signal from an electrochemical strip is sent directly to a smartphone via an audio jack. Further example is discussed for smartphone-based microfluidic assay in Chapter 7, Section 4.

These days, wireless headphones have also become available, for example, AirPod. As it utilizes Bluetooth for wireless connection, this platform enables wireless data transmission without the need for a separate A/D converter chip.

References

[1] Yoon J-Y. Introduction to biosensors: from electric circuits to immunosensors. 2nd ed. New York, NY: Springer; 2016.

[2] Christian GD, Dasgupta PK, Schug KA. Analytical chemistry. 7th ed. Hoboken, NJ: John Wiley & Sons; 2013.

[3] Skoog DA, Holler FJ, Crouch SR. Principles of instrumental analysis. 7th ed. Boston, MA: Cengage Learning; 2017.

[4] Cole-Parmer. Cole-parmer® Sourcebook. Vernon Hills, IL. 2019. p. 534. Available at: https://pim-resources.coleparmer.com/catalog-page/gh3-0534.pdf.

[5] Barnes EO, Lewis GEM, Dale SEC, Marken F, Compton RG. Generator-collector double electrode systems: a review. Analyst 2012;137:1068−81.

[6] SparkFun Electronics. Arduino uno − R3. 2013. Available at: https://commons. wikimedia.org/wiki/File:Arduino_Uno_-_R3.jpg.

[7] Halfacree G. Raspberry Pi 3 B+. 2018. Available at: https://commons.wikimedia.org/ wiki/File:Raspberry_Pi_3_B%2B_(39906369025).png.

[8] Lillehoj PB, Huang M-C, Truong N, Ho C-M. Rapid electrochemical detection on a mobile phone. Lab Chip 2013;13:2950−5.

[9] Yao Y, Jiang C, Ping J. Flexible freestanding graphene paper-based potentiometric enzymatic aptasensor for ultrasensitive wireless detection of kanamycin. Biosens Bioelectron 2019;123:178−84.

[10] Sun A, Wambach T, Venkatesh AG, Hall DA. A low-cost smartphone-based electrochemical biosensor for pont-of-care diagnostics. 2014 IEEE Biomed Circuits Syst Conf Proc 2014:312−5.

[11] Sun AC, Yao C, Venkatesh AG, Hall DA. An efficient power harvesting mobile phone-based electrochemical biosensor for pont-of-care health monitoring. Sens Actuators B Chem 2016;235:126−35.

Smartphone for glucose monitoring

Han Zhang[1], Wei Zhang[1], Anhong Zhou, PhD[2]

Department of Biological Engineering, Utah State University, Logan, UT, United States[1];
Professor, Department of Biological Engineering, Utah State University, Logan, UT, United States[2]

1. Background

According to the International Diabetes Federation (IDF), by 2017, the number of patients with diabetes has increased from 108 million in 1980 to 425 million. An estimated four million deaths were caused by diabetes. More than approximately 1.1 million children were living with type 1 diabetes, and 352 million people were at risk of developing type 2 diabetes. The medical costs related to diabetes had exceeded $700 billion worldwide, which was 12% of total spending on adults [1].

Human blood-glucose concentrations are typically in the range of 3.9–7.1 mM (70 130mg/dL) in plasma for healthy individuals, increasing to up to 40 mM (720 mg/dL) in diabetics after food intake [2]. Hyperglycemia is the technical term for high blood glucose. Blood glucose level above 130 mg/dL during fasting or above 180 mg/dl 1–2 h after eating can be defined as hyperglycemia. The hyperglycemia can lead to a wide variety of serious consequences including nerve damage, cardiovascular damages, eye problems and kidney damage. Many different types of sensors have been developed for the monitoring of glucose in blood. The enzymatic-based colorimetric and electrochemical sensors have been extensively studied during last decades [2–8]. A great insightful review about nonenzymatic electrochemical glucose sensing approaches was provided by Toghill and Compton in 2010 [7]. Spectroscopic methods such as Raman and infrared spectroscopy for noninvasive glucose detection have also be investigated by many research groups [9–11]. This chapter focuses solely on the colorimetric and electrochemical detection methods for glucose monitoring on smartphone platform.

By 2016, there are approximately 2.1 billion smartphone users worldwide, and will grow to around 2.87 billion in 2020 [12]. With billions of smartphone users, these devices have become an inseparable part of daily life, and hence integrating a blood glucose meter within a smartphone has many advantages. First, smartphone is of small size and portable, it can be held in a hand or stored in a pocket, allowing easy access and use at the point of care. Second, smartphone combines both excellent computing power and high-resolution cameras features in a single device. It can be used as the biosensor signal reader without need of any extra equipment. Finally, due to its high popularity, implementable applications, and no need of extra devices, smartphone-based diagnostics is low cost compared to traditional methods. The use of smartphone for glucose monitoring can serve as an effective tool in controlling/reducing the progression of diabetes and subsequently improving quality of life.

2. Role of smartphone for glucose monitoring

Smartphone-based measurement applications were found in interdisciplinary fields, including but not limited to medical diagnosis, environmental monitoring, agriculture, and biological sensing. Glucose monitoring is one of the most significant fields of smartphone-based measurement. As colorimetric and electrochemical methods are the two most frequently applied techniques for glucose measurement, in this chapter, we will focus on these two methods.

For both colorimetric and electrochemical methods, the functions of smartphone include the modules of signal acquisition, signal processing and result display. The smartphone application is developed to commend smartphone (Scheme 4.1). For colorimetric measurement, image of sensing zone is captured by smartphone camera, then, the captured image is processed to certain image format. For example, the image is first converted to a grayscale image to obtain the average gray value of the sensing zone. After that, the gray value is substituted into the calibration equation to obtain a concentration of analyte. For electrochemical measurement, smartphone collects electrical signal from electron transport of the chemical reactions on biosensor. Then, the signal is processed by the microcircuit board and output electrical signal is then transferred to the smartphone. Finally, the built-in application analyzes the signal and calculates the glucose concentration. The major hardware difference between colorimetric and electrochemical methods is that the electrochemical method generally has a mini-potentiostat and/or a microcontroller for voltage control and digital-to-analog conversion. Examples of the smartphone-

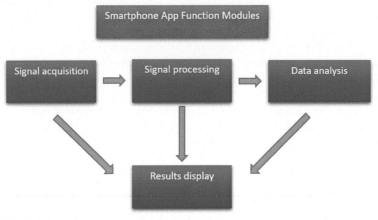

SCHEME 4.1

The Android application function modules.

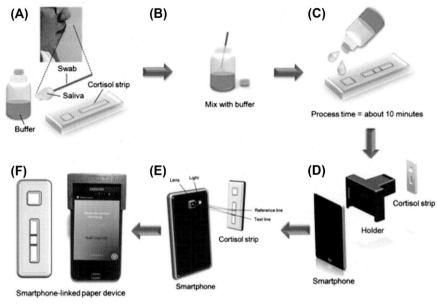

FIGURE 4.1

A photograph and schematic diagram of the smartphone-based cortisol measurement system. (A and B) Saliva collection. (C) Buffer loading. (D) View of the system showing placement of holder and strip. (E and F) Smartphone-based reading system.

Reprinted from Ref. [13] with permission. (C) 2014 Elsevier B.V.

based colorimetric and electrochemical monitoring are shown in Fig. 4.1 and Fig. 4.2, respectively. Fig. 4.1 shows a smartphone real-time monitoring of human salivary cortisol with colorimetric method, and Fig. 4.2 illustrates a smartphone-based cyclic voltammetry (CV) method for detection of blood glucose.

3. Colorimetric method

In colorimetric glucose detection, glucose oxidase (GOx) and horseradish peroxidase (HRP) is a common enzyme pair used to catalyze the reaction between glucose and the chromogenic substrates [5,15-17]. The GOx first oxidizes glucose to gluconic acid and hydrogen peroxide (H_2O_2). Then, HRP catalyzes the reaction of H_2O_2 with chromogenic substrates and exhibits a blue color. The commonly used chromogenic substrates for HRP include potassium iodide, 3,3',5,5'-tetramethylbenzidine, 3,3'-diaminobenzidine, 2,2'-azino-bis(3-ethylbenzothiazoline-6-sulphonic acid), and others [18]. The resulting color intensity is proportional to the concentration of analyte. To quantify the analytes, a digital camera, cell phone, or scanner is typically used to record the color intensity [19]. In the colorimetric analysis, color information could be obtained with paper-based sensors to quantify the color

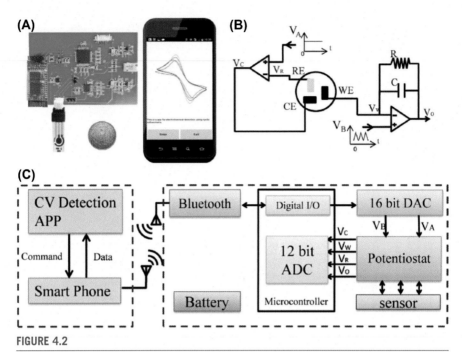

FIGURE 4.2

The schematic and the design of the smartphone-based cyclic voltammetry (CV) system.
(A) The image of the handheld detector connected with SPE. (B) The circuit design of the
potentiostat. (C) A schematic diagram of the smartphone-based CV system.

Reprinted from Ref. [14] with permission. (C) 2017 Elsevier B.V.

variation in different image format such as grayscale, RGB, and HSV (hue, saturation, and lightness), which is an alternative representation of the RGB color model [20-24].

Due to its global affordability, user-friendliness, portability, and rapid visual readout, colorimetric detection could be the very attractive and convenient technique to be coupled with lateral flow assays (LFAs) for biomedical diagnostics. The lateral flow assay is a paper-based platform for the detection and quantification of analytes in complex mixtures, where the sample is placed on a test device and the results are displayed within a few minutes. Microfluidic paper-based device (μPAD) combined with LFA for glucose detection was frequently implemented as the sensing platform for glucose colorimetric assays, which was first proposed by the Whitesides group in 2007 (Fig. 4.3) [25]. Their device was fabricated by paper-based photolithography with a measurement range of glucose concentration from 0 to 500 mM. Whatman chromatography paper was the most commonly used lateral flow paper substrate [5,15,26]. The hydrophobic barriers were constructed with many techniques such as wax printing, photolithography, hot embossing, and screen printing [18]. After that, the enzymes and substrates are loaded onto the device. The limit of detection

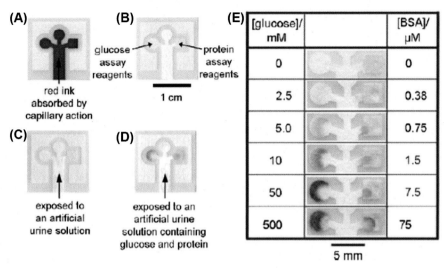

FIGURE 4.3

Whatman chromatography paper patterned with photoresist. The darker lines are cured photoresist, whereas the lighter areas are unexposed paper. The paper-based device was evaluated with glucose and protein assay.

Reprinted from reference [25] with permission. (C) 2007 Wiley-VCH.

(LOD) with image data collected by a scanner or common digital camera is often reported in the literature as ranging from 0.1 to 5 mM (1.8mg/dL to 90mg/dL) [18,27].

3.1 Challenges of smartphone-based colorimetric method

Although smartphone-based colorimetric detection provides many advantages, it also faces some drawbacks that can adversely affect the analytical performance of µPAD. Variations of image acquisition conditions and uneven color development in the detection area of the µPAD are two major problems. The change of image capture condition would apparently influence the quality of image acquisition. In general, quantitative analysis with colorimetry on µPADs requires the following steps: the capture of images, the processing of images, and the construction of a calibration curve. The most important issue for precise and accurate onsite measurements is to maintain the constant image acquisition conditions (such as ambient light intensity, exposure time, white balance, and light sensitivity of implemented camera), as the color intensity of an image is sensitive to the spectrum and intensity of the ambient light source. The uneven manual reagent loading and the spreading of the fluid in porous cellulose fiber structure can cause uneven color development in the detection area. This is particularly problematic when the colored product concentrates at the edge of the detection zone [28]. These challenges are further explained in the following sections.

3.2 Uneven color development

The uneven color development was commonly reported in the published literatures [29-32]. The problem of uneven color development includes washing effect, nonuniform intensity formed in the detection zones, and enzyme/substrate loading overflow. The washing effect is caused by the washing of the reagents and enzymes from the printed position to the edges of the testing zone when the sample solution travels through the patterned channels. Nonuniform intensity is caused by the mobility of the fluid in porous cellulose fiber structure, and overflowing is primarily caused by unbalanced manual reagent loading. The examples of the uneven color development are illustrated in Fig. 4.4. The poor control of manual reagent loading and the spreading of the fluid in porous cellulose fiber structure are the major causes of uneven color development in the detection area.

The uneven color development issue can be addressed by introduction of inkjet printing method being developed in our research laboratory, instead of using manual loading. The Fuji DMP 2831 material printer was employed in this case. The nominal drop size of the material printer is 10 pL. Generally, a 50 μm diameter spot can be produced on chromatography paper substrate. Therefore, the designed pattern could be printed out precisely and without overflowing problem because of the high printing resolution, and the spreading of liquid ink can also be minimized due to the extra low volume of each ink drop. Therefore, the overflowing and uneven color development issue can be addressed. Fig. 4.5A compares the results from both manual pipetting and printing method. As we can see, the printed μPAD had better color uniformity. Another issue is the washing effect. The application of chitosan (CHI) as immobilization support has been successfully demonstrated by other researchers for sensing studies [5,33,34]. Here, the CHI was

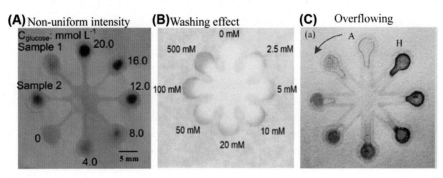

FIGURE 4.4

Recently reported cases of uneven color development. non-uniform intensity (a) washing effect (b) and overflowing (c) on paper based colorimetric test strips

Reprinted from reference [30] with permission; (C) 2015 Royal Society of Chemistry, Reprinted from reference [29] with permission; (C) 2014 Royal Society of Chemistry, Reprinted from reference [31] with permission; (C) 2016 Royal Society of Chemistry.

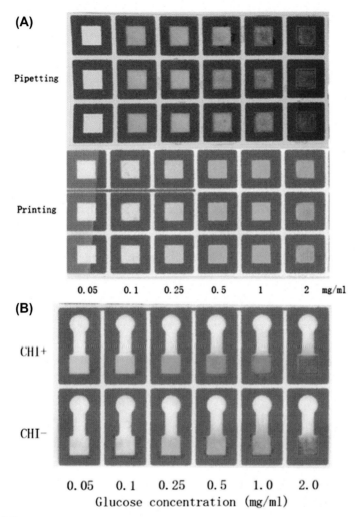

FIGURE 4.5

Comparison of μPADs with substrate and enzymes loaded with pipetting (top) and printing method (bottom), printed with a Fuji DMP 2831 printer (A). Images showing the inkjet-printed colorimetric performance of μPADs and without and with chitosan (B).

inkjet-printed to immobilize the enzymes (GOx and HRP) by the entrapment method [35] to minimize the washing effect. Fig. 4.5B shows the comparison of inkjet-printed colorimetric performance of μPADs without and with chitosan coating. According to the result illustrated in Fig. 4.5B, the cyan colors in the sensing zones of CHI+ were uniformly distributed without washing effect and overflowing of color.

3.3 Effect of ambient light

Variation of ambient light significantly affects the background intensity of an image. Many colorimetric sensor measurement studies have focused on finding the true color of a material by isolating the background noise that may be caused by ambient light. Fig. 4.6 compares the colorimetric test results from both the smartphone-based method and traditional scanner method. Glucose concentrations from 0.05 to 1 mg/ml (5–100 mg/dL) were used for this evaluation. The images of the same sample were captured from both indoor and outdoor environment, and compared with scanned result. According to the result, the scanner method had the best linear regression ($R^2 = 0.9979$), due to illumination conditions were the same at all locations during scanning. The standard deviations of the smartphone-based method were much larger than the scanner method, and this may be caused by variation of the angle and distance during the image capture. More importantly, the major difference of grayscale value at the same concentration point was caused by the variance of ambient light intensity in different environments. Given the complication of the ambient light source, recalibration is needed if the test location is changed. The test strips with consistent illumination would significantly improve the performance of the method.

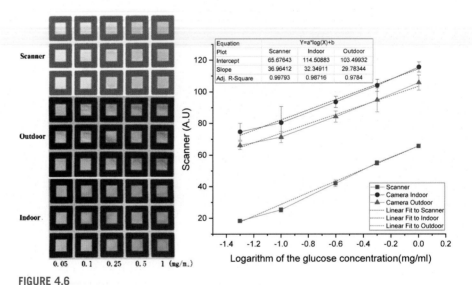

FIGURE 4.6

Paper-based device images collected under different methods and environment (left), and glucose analytical curves of different collection methods (all collection methods used the same sample of paper).

3.4 Effect of exposure time and light sensitivity

Another major challenge of using a smartphone for colorimetric readout is the variation of the image capture parameters on different brands of smartphones while using a built-in automatic camera application. Parameters such as exposure time, light sensitivity, and white balance would significantly affect the signal baseline of images. Fig. 4.7A demonstrates two different gray values for the same µPAD sample placed in the optical chamber. As the sample was placed inside an optical chamber, the ambient light conditions was the same to both images. A built-in automatic camera application was used for this image acquisition. The camera parameters automatically change every capture, and as a result, the gray intensity of the same area varies

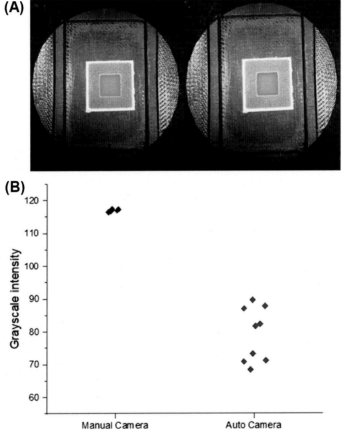

FIGURE 4.7

Effect of exposure using two different types of camera applications on Samsung Galaxy J3 smartphone. (A) The gray intensity difference of two exposures of the same sample at auto-exposure condition. (B) The distribution of detected gray value for the same sample with different camera applications.

significantly for each image capture. To minimize these effects, different camera application is needed with the ability to manually set and fix these parameters at a certain value during the image capture. As seen in Fig. 4.7B, the gray intensities were much stable while all related parameters were fixed using a manual camera application for image capture.

3.5 3D-printed enclosure

As we discuss earlier, one major challenge in the process of measuring colorimetric sensor is isolating ambient light effects. Ambient light conditions are so complex that simple software algorithms cannot isolate them effectively. To minimized the effect, optical chambers and enclosures were introduced by various researchers [36,37]. Fig. 4.8A shows an example of optical enclosures to eliminate ambient light effect. The 3D-printed enclosure was printed to block the ambient light using plastic

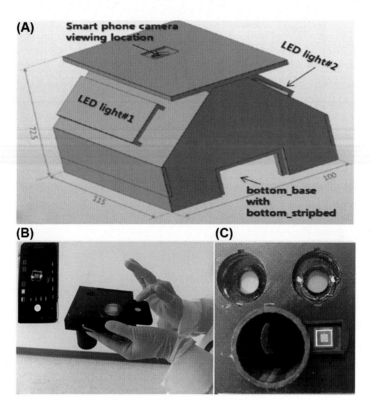

FIGURE 4.8

Examples of optical enclosure for eliminating ambient light. 3D-printed mini light box; the unit is mm (a). Reprinted from reference [36] with permission. (C) 2017 S.D. Kim, Y. Koo and Y. Yun. 3D-printed platform with a smartphone for detection (b) and bottom view of the platform with the paper-based device (c).

materials such as acrylonitrile butadiene styrene. The LED lights were inserted and fixed inside of the enclosure. As the amount and voltage of LED light is constant, the illumination background is hence consistent. Fig. 4.8(B and C) illustrates a complete smartphone analytical platform developed by our research laboratory. The platform was printed with VeroBlackPlus black materials with Objet260 Connex3 3D printer. The LEDs were placed and fixed in the ring structure under the smartphone holder. The LED light voltage was set to 3 V, and the light reflects from bottom of the optical chamber and is collected by smartphone camera. The μPAD can be inserted from both sides of the chip holder, and only the sensing zone is exposed as valid signal collection area.

3.6 Blood filtration

Due to difficulty on signal readout caused by the chemical complexity of the matrix and the natural red background color of hemoglobin, detection of analytes in whole blood with smartphone colorimetric method may not be achieved directly. Most biological samples require steps to remove cells from blood because cells often interfere with the analytical measurements [38]. As a result, biochemical tests are typically carried out in serum or plasma [39]. The paper-based blood separation method can be separated into two categories including the lateral separation method and vertical separation method. The lateral separation: blood cells are separated from plasma during the flowing phase due to the difference of flow speed of small molecules and cells. Various types of papers have been demonstrated with this method. For example, VF1, VF2, and LF1 filter papers from Whatman were frequently used as the materials for a lateral flow separation method [6,40,41]. Recent study from the literature indicated that LF1 membrane was the most suitable for blood separations for μPAD application (Fig. 4.9A) [40]. The vertical separation means that the blood cells and plasma are separated by vertical flowing. The commercially available filter membrane such as Vivid (PALL, New York) blood separation membrane GF, GX, and GR from Pall were reported being applied for vertical flow separation method. These membranes are usually unidirectional filtration membranes [42,43]. Examples of both lateral and vertical flow separation methods discussed are shown in Fig. 4.9.

4. Electrochemical method

Electrochemical sensors are attractive due to their fast time-response, straightforward for operation, and the relatively high sensitivity of electrochemical measurements [45]. A number of electrochemical techniques have been applied for detection of glucose, including potentiometry, coulometry and amperometry. The historical commencement of biosensors was in 1960s with the pioneering work of Clark and Lyons, and the first enzyme-based glucose sensor commenced by Updike and Hicks in 1967 [46]. Since then, electrochemical methods, especially

(A)

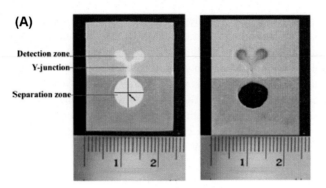

(B)

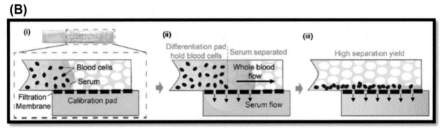

FIGURE 4.9

Paper-based blood separation. (A) Lateral flow separation, μPAD applied with whole blood concurrence with plasma separation and determination of human serum protein. (B) Vertical flow separation, illustration of the flow dynamics of the applied platform. The platform slows down the fouling layer aggregation by holding blood cells upstream, and thus the separation process becomes faster and more efficient.

(A) Reprinted from Ref. [41] with permission. (C) 2012 Royal Society of Chemistry. (B) Reprinted from Ref. [44] with permission. (C) 2018 Royal Society of Chemistry.

amperometric methods, have been widely utilized in glucose sensing based on glucose oxidase (GOx). The Clark–Lyons demonstration of a sensor based on the reaction between glucose and glucose oxidase uses the reaction below with potentiometric electrochemistry for the signal transduction. Although reported nearly half century ago, this basic sensing formula remains the favorable explanation of sensing mechanism today.

$$GluOx(FAD) + Glucose \rightarrow GluOx(FADH_2) + Glucono\text{-}D\text{-}lactone$$

$$GluOx(FADH_2) + O_2 \rightarrow GluOx(FAD) + H_2O_2$$

Due to its rapid electron transfer rate, the use of pyrroloquinoline quinone (PQQ) as a cofactor to GDH has become the gold standard of the commercial glucose sensing [47].

$$GDH\text{-}PQQ \text{ (oxidized)} + Glucose \rightarrow GDH\text{-}PQQ \text{ (reduced)} + Glucono\text{-}D\text{-}lactone$$

Similar to FAD$^+$, the reduced PQQ will be oxidized back to its oxidized form, a change in current can be observed with application of constant potential.

4.1 Generations of electrochemical glucose sensor

Three generations of glucose biosensors based on different mechanisms of electron transfer, including the use of natural secondary substrates, artificial redox mediators, or direct electron transfer, are summarized in Fig. 4.10. In the first-generation glucose biosensors, glucose was decomposed catalytically via GOx and the H$_2$O$_2$ was generated and then subsequently oxidized at the electrode surface, producing a measurable current signal. The first-generation glucose biosensor relies on the use of the oxygen as the physiological electron acceptor. They are subject to errors resulting from depletion in local oxygen concentration. The second-generation of glucose biosensors have been proposed for addressing this oxygen limitation. Artificial electron mediators (M), for example, ferro/ferricyanide, hydroquinone, ferrocene, and various redox organic dyes between the electrode and the GOx are employed. These mediators make the electron transfer rate between the electrode and the GOx faster and also give a way of getting around for a case when limited oxygen pressure commonly observed from the first-generation glucose sensor [32]. In the third-generation glucose biosensor, the GOx is directly coupled to the electrode. The direct electron transfer efficiently generates an amperometric output

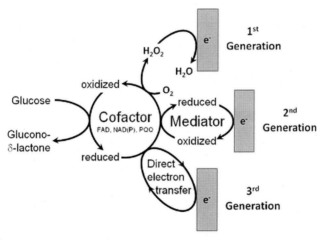

FIGURE 4.10

Schematic representation of the principles of first-, second-, and third-generation glucose sensors. Electrons from the glucose oxidation reaction are first taken up by the enzyme's cofactor (primary electron acceptor) and transferred to either oxygen (first generation), an electron mediator (second generation), or directly to the electrode (third generation).

Reprinted from Ref. [3] with permission. (C) 2011 SAGE Publishing.

signal. The improved sensing performance by the direct electron transfer has been realized by incorporating the enzyme with metal nanoparticles and semiconductive nanomaterials [4].

4.2 Smartphone-based electrochemical sensor device

In a smartphone-based electrochemical sensor device, sensor, transducer, and smartphone are integrated together. Fig. 4.11 illustrates a smartphone-based electrochemical biosensor for point-of-care diagnoses. It consists of three parts: the electrodes system (chip) where chemical reaction occurs, the adaptor contains microcontroller or miniaturized potentiostat for signal collection, as well as the smartphone application to analyze and display the result.

The key factors impacting the integration of electrochemical glucose sensors into smartphone platforms are the accuracy and high cost. Until now, the accuracy of smartphone-based glucose monitoring had been so far inferior to those of blood glucose meters for measurement of capillary blood glucose. Although the performance of this method has been improved significantly, accuracy is still strongly dependent on the glucose level, especially at the low glucose concentration level [48]. The cost is another issue, and the price of the smartphone adaptor commonly ranges from tens to hundreds of US dollars [8]. Commercial glucose meters, on the other hand, cost much less (for most brands, the price is lower than $30 USD and comes for free if purchase over 100 strips). Some glucose meters have built-in Bluetooth feature. Besides, the enzymatic-functionalized electrodes are more expensive than the paper-based colorimetric strips. They are destroyed after only a few uses due to gradual leaching of sensor reagents. Cost becomes a major obstacle prohibiting its use in those resource-limited settings. Although development of reusable sensors has been reported by several groups, their products still take time to be commercialized [49-51].

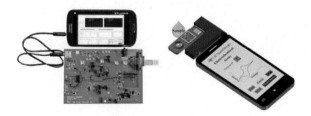

FIGURE 4.11

Smartphone-based glucose measurement device. Photograph of prototype board powered by a Samsung Galaxy S3 (left). 3D rendered prototype of proposed system (right) [52].

Reprinted from Ref. [52] with permission. (C) 2014 IEEE.

5. Noninvasive glucose monitoring

Noninvasive glucose sensing is the ultimate goal of glucose monitoring. Diabetic patient must regularly measure their blood glucose concentration and to decide how much medication to use. Currently, this blood glucose monitoring is generally completed at home using an invasive method (finger prick-based blood glucose meter). Although relatively high accuracy than noninvasive method, this invasive pricking test makes some patients painful and uncomfortable, as a result, many patients are not monitoring as frequently as they are supposed to be. Researchers have performed numerous studies regarding noninvasive blood glucose monitoring [53-55]. However, currently, noninvasive glucose monitoring does not satisfy our demand due to its uncertain measurement accuracy and weak correlation with the human blood plasma glucose concentration.

Optical spectroscopic methods have been attempted as alternatives, which were not very successful. Infrared spectrophotometry are easily influenced by interference from background substrate [56]. Impedance spectroscopy and Raman spectroscopy lack specificity to glucose [57,58]. Direct glucose monitoring from easily obtainable body fluids such as sweat, saliva, and tear has also been suggested, while they were not accurate since glucose is 10 to 100 times diluted in those fluids [59-61]. Although a simple, pain-free, noninvasive method would make significant improvement in diabetes care, it still needs significant further study and optimization in sensor design.

6. Regulation, requirement and constraint

Many clinical facilities use handheld glucose meters, which provide rapid quantitation of glucose using small volumes (less than 1 μL) of whole blood. The minimal sample volume required, rapid analysis time, ease of use, and portability make these glucose meters an ideal tool to help manage blood glucose level. Thus, glucose meters are used throughout hospitals, in doctor's office, laboratories, operation theater, and intensive care units [62]. Glucose meter can also be used at home to determine daily adjustment in treatment, to monitor dangerously high or low levels of glucose, and to understand how diet and exercise affect the glucose level [63]. Because of its extensive use in various clinical scenarios, regulatory specifications will remain a challenge [64].

6.1 Accuracy requirement

In 2016, the U.S. Food and Drug Administration (FDA) finalized its recommendations ("guidance") on accuracy and other requirements for new blood glucose meters, which did not impact meters and strips already on the market. For people with diabetes, the main outcomes of the new guidance are that *new* meters will need to be more accurate, conduct larger studies before they are approved, and more clearly display accuracy on their packaging [65].

For glucose meters used by people with diabetes at home, the FDA guidance specifies the following accuracy standards:

- 95% of all measured blood glucose meter values must be within 15% of the true value (a laboratory measurement); and
- 99% of meter values must be within 20% of the true value.

Comparatively, the prior rules called for 20% accuracy for most blood sugar ranges. The new accuracy standards indicate if the true (lab measured) glucose value is 100 mg/dl that means the meter has to be within ±15 mg/dl (85−115 mg/dl) in 95% of cases, and within ±20 mg/dl (80−120 mg/dl) in 99% of cases. Studies of new meters must also include at least 350 people with diabetes, larger than previously required.

For healthcare providers using glucose meters in facilities ("point-of-care testing"), a more tightened guidance has been issued:

- 95% of meter values should be within 12% of the reference value for blood sugars over 75 mg/dl, and within 12 mg/dl for blood sugars below 75 mg/dl; and
- 98% of meter values should be within 15% of the reference value for blood sugars over 75 mg/dL, and within 15 mg/dl for blood sugars below 75 mg/dl.

6.2 Other requirements

Manufacturing Process: Beyond the accuracy standard, the FDA also cracked down on manufacturers' lot release methodology by data collection and site inspections to supervise the production process and ensure the quality of their production.

Labeling: Most importantly, FDA has called for new labeling information on test strip vials; they must include information about the lot/production, and a description of the performance (accuracy data) on the outer box label, so users can compare 1 m to another.

Third-Party Test Strips: Importantly, the final rules add provisions specific to off-brand test strip makers that have come under criticism in recent years. Although these strips are often less-expensive, they haven't been subject to the same requirements for accuracy as the meters—particularly when certain brands are made overseas, and the FDA isn't able to inspect manufacturing facilities like they do in the United States. Now, FDA rules say these manufacturers should "ensure that they are aware of any design changes to the meter because such changes could affect compatibility of the test strip with the meter." This must be addressed in the third-party company's 510(k) filing, and the FDA also recommends them to submit the agreement documentation between the third-party strip maker and meter manufacturer.

FDA has issued the guidance for mobile medical applications in 2013, which explains definition of mobile medical apps and provides examples of how the FDA might regulate certain moderate-risk (Class II) and high-risk (Class III) mobile medical apps [66].

Conclusion

Smartphones are the most promising and effective means of delivering mobile healthcare by enabling better personal health monitoring and rapid diagnostics. In this chapter, smartphone-based colorimetric and electrochemical methods, as two major methods for glucose monitoring, were discussed. First, the basic functional structure of combination glucose measurement with smartphone platform was described. Then, the general mechanism of both colorimetric and electrochemical detection of glucose was described. After that, the challenges of each of these methods themselves as well as combination with smartphone were proposed and possible solutions for both methods were explored. In addition, paper-based blood filtration and noninvasive glucose monitoring were reviewed. Finally, the prospective of regulation, requirement, and constraints was highlighted. Overall, the development of a low-cost, high-accuracy, and noninvasive measurement method is the future goal of smartphone-based self-monitoring of blood glucose.

Acknowledgment

We appreciate the funding support from Utah Water Research Laboratory and Utah Agricultural Experiment Station.

References

[1] Diabetes facts & figures. https://www.idf.org/aboutdiabetes/what-is-diabetes/facts-figures.html (accessed 05/22/2019).

[2] Bruen D, Delaney C, Florea L, Diamond D. Glucose sensing for diabetes monitoring: recent developments. Sensors 2017;17(8):1866.

[3] Ferri S, Kojima K, Sode K. Review of glucose oxidases and glucose dehydrogenases: a bird's eye view of glucose sensing enzymes. SAGE Publications; 2011.

[4] Rahman M, Ahammad A, Jin J-H, Ahn SJ, Lee J-J. A comprehensive review of glucose biosensors based on nanostructured metal-oxides. Sensors 2010;10(5):4855−86.

[5] Gabriel EF, Garcia PT, Cardoso TM, Lopes FM, Martins FT, Coltro WK. Highly sensitive colorimetric detection of glucose and uric acid in biological fluids using chitosan-modified paper microfluidic devices. Analyst 2016;141(15):4749−56.

[6] Noiphung J, Songjaroen T, Dungchai W, Henry CS, Chailapakul O, Laiwattanapaisal W. Electrochemical detection of glucose from whole blood using paper-based microfluidic devices. Anal Chim Acta 2013;788:39−45.

[7] Toghill KE, Compton RG. Electrochemical non-enzymatic glucose sensors: a perspective and an evaluation. Int J Electrochem Sci 2010;5(9):1246−301.

[8] Vashist S, Schneider E, Luong J. Commercial smartphone-based devices and smart applications for personalized healthcare monitoring and management. Diagnostics 2014; 4(3):104−28.

[9] Yadav J, Rani A, Singh V, Murari BM. Prospects and limitations of non-invasive blood glucose monitoring using near-infrared spectroscopy. Biomed Signal Process Control 2015;18:214—27.

[10] Pandey R, Paidi SK, Valdez TA, Zhang C, Spegazzini N, Dasari RR, Barman I. Noninvasive monitoring of blood glucose with Raman spectroscopy. Acc Chem Res 2017; 50(2):264—72.

[11] Spegazzini N, Barman I, Dingari NC, Pandey R, Soares JS, Ozaki Y, Dasari RR. Spectroscopic approach for dynamic bioanalyte tracking with minimal concentration information. Sci Rep 2014;4:7013.

[12] Number of smartphone users worldwide from 2014 to 2020 (in billions). https://www.statista.com/statistics/330695/number-of-smartphone-users-worldwide/ (accessed 05/22).

[13] Choi S, Kim S, Yang J-S, Lee J-H, Joo C, Jung H-I. Real-time measurement of human salivary cortisol for the assessment of psychological stress using a smartphone. Sensing and Bio-Sensing Research 2014;2:8—11.

[14] Ji D, Liu L, Li S, Chen C, Lu Y, Wu J, Liu Q. Smartphone-based cyclic voltammetry system with graphene modified screen printed electrodes for glucose detection. Biosens Bioelectron 2017;98:449—56.

[15] Gabriel E, Garcia P, Lopes F, Coltro W. Paper-based colorimetric biosensor for tear glucose measurements. Micromachines 2017;8(4):104.

[16] Figueredo F, Garcia PT, Cortón E, Coltro WKT. Enhanced analytical performance of paper microfluidic devices by using Fe_3O_4 nanoparticles, MWCNT, and graphene oxide. ACS Appl Mater Interfaces 2015;8(1):11—5.

[17] Evans E, Gabriel EF, Benavidez TE, Tomazelli Coltro W, Garcia CD. Modification of microfluidic paper-based devices with silica nanoparticles. Analyst 2014;139(21): 5560—7.

[18] Liu S, Su W, Ding X. A review on microfluidic paper-based analytical devices for glucose detection. Sensors 2016;16(12).

[19] Martinez AW, Phillips ST, Whitesides GM, Carrilho E. Diagnostics for the developing world: microfluidic paper-based analytical devices. ACS Publications; 2009.

[20] Bayram A, Horzum N, Metin AU, Kılıç V, Solmaz ME. Colorimetric bisphenol-A detection with a portable smartphone-based spectrometer. IEEE Sens J 2018.

[21] Jia M-Y, Wu Q-S, Li H, Zhang Y, Guan Y-F, Feng L. The calibration of cellphone camera-based colorimetric sensor array and its application in the determination of glucose in urine. Biosens Bioelectron 2015;74:1029—37.

[22] Morsy MK, Zór K, Kostesha N, Alstrøm TS, Heiskanen A, El-Tanahi H, Sharoba A, Papkovsky D, Larsen J, Khalaf H. Development and validation of a colorimetric sensor array for fish spoilage monitoring. Food Control 2016;60:346—52.

[23] Lopez-Ruiz N, Curto VF, Erenas MM, Benito-Lopez F, Diamond D, Palma AJ, Capitan-Vallvey LF. Smartphone-based simultaneous pH and nitrite colorimetric determination for paper microfluidic devices. Anal Chem 2014;86(19):9554—62.

[24] Mutlu AY, Kılıç V, Özdemir GK, Bayram A, Horzum N, Solmaz ME. Smartphone-based colorimetric detection via machine learning. Analyst 2017;142(13):2434—41.

[25] Martinez AW, Phillips ST, Butte MJ, Whitesides GM. Patterned paper as a platform for inexpensive, low-volume, portable bioassays. Angew Chem 2007;46(8):1318—20.

[26] Zhu W-J, Feng D-Q, Chen M, Chen Z-D, Zhu R, Fang H-L, Wang W. Bienzyme colorimetric detection of glucose with self-calibration based on tree-shaped paper strip. Sens Actuators B Chem 2014;190:414—8.

[27] Martinez AW, Phillips ST, Carrilho E, Thomas III SW, Sindi H, Whitesides GM. Simple telemedicine for developing regions: camera phones and paper-based microfluidic devices for real-time, off-site diagnosis. Anal Chem 2008;80(10):3699−707.

[28] Li X, Tian J, Shen W. Quantitative biomarker assay with microfluidic paper-based analytical devices. Anal Bioanal Chem 2010;396(1):495−501.

[29] Mohammadi S, Maeki M, Mohamadi RM, Ishida A, Tani H, Tokeshi M. An instrument-free, screen-printed paper microfluidic device that enables bio and chemical sensing. Analyst 2015;140(19):6493−9.

[30] Cai L, Wang Y, Wu Y, Xu C, Zhong M, Lai H, Huang J. Fabrication of a microfluidic paper-based analytical device by silanization of filter cellulose using a paper mask for glucose assay. Analyst 2014;139(18):4593−8.

[31] Yang J, Zhu L, Tang W. Fabrication of paper micro-devices with wax jetting. RSC Adv 2016;6(22):17921−8.

[32] Cass AE, Davis G, Francis GD, Hill HAO, Aston WJ, Higgins IJ, Plotkin EV, Scott LD, Turner AP. Ferrocene-mediated enzyme electrode for amperometric determination of glucose. Anal Chem 1984;56(4):667−71.

[33] Vosmanská V, Kolářová K, Rimpelová S, Kolská Z, Švorčík V. Antibacterial wound dressing: plasma treatment effect on chitosan impregnation and in situ synthesis of silver chloride on cellulose surface. RSC Adv 2015;5(23):17690−9.

[34] Liu W, Cassano CL, Xu X, Fan ZH. Laminated paper-based analytical devices (LPAD) with origami-enabled chemiluminescence immunoassay for cotinine detection in mouse serum. Anal Chem 2013;85(21):10270−6.

[35] Witkowska Nery E. Analysis of glucose, cholesterol and uric acid. In: Analysis of samples of clinical and alimentary interest with paper-based devices; 2016. p. 25−108.

[36] Kim SD, Koo Y, Yun Y. A smartphone-based automatic measurement method for colorimetric pH detection using a color adaptation algorithm. Sensors 2017;17(7):1604.

[37] Chun HJ, Park YM, Han YD, Jang YH, Yoon HC. Based glucose biosensing system utilizing a smartphone as a signal reader. BioChip J 2014;8(3):218−26.

[38] Crowley TA, Pizziconi V. Isolation of plasma from whole blood using planar microfilters for lab-on-a-chip applications. Lab Chip 2005;5(9):922−9.

[39] VanDelinder V, Groisman A. Separation of plasma from whole human blood in a continuous cross-flow in a molded microfluidic device. Anal Chem 2006;78(11):3765−71.

[40] Yang X, Forouzan O, Brown TP, Shevkoplyas SS. Integrated separation of blood plasma from whole blood for microfluidic paper-based analytical devices. Lab Chip 2012;12(2):274−80.

[41] Songjaroen T, Dungchai W, Chailapakul O, Henry CS, Laiwattanapaisal W. Blood separation on microfluidic paper-based analytical devices. Lab Chip 2012;12(18):3392−8.

[42] Liu C, Mauk M, Gross R, Bushman FD, Edelstein PH, Collman RG, Bau HH. Membrane-based, sedimentation-assisted plasma separator for point-of-care applications. Anal Chem 2013;85(21):10463−70.

[43] Liu C, Liao S-C, Song J, Mauk MG, Li X, Wu G, Ge D, Greenberg RM, Yang S, Bau HH. A high-efficiency superhydrophobic plasma separator. Lab Chip 2016;16(3):553−60.

[44] Heller A. Miniature biofuel cells. Phys Chem Chem Phys 2004;6(2):209−16.

[45] Lu Z, Rey E, Vemulapati S, Srinivasan B, Mehta S, Erickson D. High-yield paper-based quantitative blood separation system. Lab Chip 2018;18(24):3865−71.

[46] Wang J. Electrochemical glucose biosensors. Chem Rev 2008;108(2):814−25.

[47] Yoon J-Y. Introduction to biosensors: from electric circuits to immunosensors. Springer; 2016.

[48] Thabit H, Hovorka R. Bridging technology and clinical practice: innovating inpatient hyperglycaemia management in non-critical care settings. Diabet Med 2018;35(4): 460−71.

[49] Bandodkar AJ, Imani S, Nunez-Flores R, Kumar R, Wang C, Mohan AV, Wang J, Mercier PP. Re-useable electrochemical glucose sensors integrated into a smartphone platform. Biosens Bioelectron 2018;101:181−7.

[50] Yang H-W, Hua M-Y, Chen S-L, Tsai R-Y. Reusable sensor based on high magnetization carboxyl-modified graphene oxide with intrinsic hydrogen peroxide catalytic activity for hydrogen peroxide and glucose detection. Biosens Bioelectron 2013;41: 172−9.

[51] Hu J, Yu Y, Brooks JC, Godwin LA, Somasundaram S, Torabinejad F, Kim J, Shannon C, Easley CJ. A reusable electrochemical proximity assay for highly selective, real-time protein quantitation in biological matrices. J Am Chem Soc 2014;136(23): 8467−74.

[52] Sun A, Wambach T, Venkatesh A, Hall DA. A low-cost smartphone-based electrochemical biosensor for point-of-care diagnostics. In: Biomedical circuits and systems conference (BioCAS), 2014 IEEE. IEEE; 2014. p. 312−5.

[53] Bandodkar AJ, Jia W, Wang J. Tattoo-based wearable electrochemical devices: a review. Electroanalysis 2015;27(3):562−72.

[54] Tura A, Maran A, Pacini G. Non-invasive glucose monitoring: assessment of technologies and devices according to quantitative criteria. Diabet Res Clin Pract 2007; 77(1):16−40.

[55] Vashist SK. Non-invasive glucose monitoring technology in diabetes management: a review. Anal Chim Acta 2012;750:16−27.

[56] Ghosn MG, Sudheendran N, Wendt M, Glasser A, Tuchin VV, Larin KV. Monitoring of glucose permeability in monkey skin in vivo using optical coherence tomography. J Biophot 2010;3(1-2):25−33.

[57] Shervedani RK, Mehrjardi AH, Zamiri N. A novel method for glucose determination based on electrochemical impedance spectroscopy using glucose oxidase self-assembled biosensor. Bioelectrochemistry 2006;69(2):201−8.

[58] Yang X, Zhang AY, Wheeler DA, Bond TC, Gu C, Li Y. Direct molecule-specific glucose detection by Raman spectroscopy based on photonic crystal fiber. Anal Bioanal Chem 2012;402(2):687−91.

[59] Gao W, Emaminejad S, Nyein HYY, Challa S, Chen K, Peck A, Fahad HM, Ota H, Shiraki H, Kiriya D. Fully integrated wearable sensor arrays for multiplexed in situ perspiration analysis. Nature 2016;529(7587):509.

[60] Soni A, Jha SK. A paper strip based non-invasive glucose biosensor for salivary analysis. Biosens Bioelectron 2015;67:763−8.

[61] Lee H, Choi TK, Lee YB, Cho HR, Ghaffari R, Wang L, Choi HJ, Chung TD, Lu N, Hyeon T. A graphene-based electrochemical device with thermoresponsive microneedles for diabetes monitoring and therapy. Nat Nanotechnol 2016;11(6):566.

[62] Isbell TS. A review of compliant glucose POCT options. Med Lab Manag 2016;5(1):2.

[63] FDA Blood Glucose Monitoring Devices. https://www.fda.gov/medicaldevices/ productsandmedicalprocedures/invitrodiagnostics/glucosetestingdevices/default.htm.

[64] Tirimacco R, Koumantakis G, Erasmus R, Mosca A, Sandberg S, Watson ID, Goldsmith B, Gillery P, International Federation of Clinical C. Laboratory medicine

working group on glucose point-of-care, T., glucose meters — fit for clinical purpose. Clin Chem Lab Med 2013;51(5):943—52.

[65] Healthline FDA updates guidelines on glucose meter accuracy. https://www.healthline.com/diabetesmine/fda-finalizes-meter-accuracy-rules#1.

[66] Mobile medical applications. https://www.fda.gov/media/80958/download (accessed 05/22/2019).

Smartphone-based flow cytometry

Zheng Li, PhD [1], Shengwei Zhang, [1], Qingshan Wei, PhD [2]

[1]*Department of Chemical and Biomolecular Engineering, North Carolina State University, Raleigh, NC, United States;* [2]*Assistant Professor, Department of Chemical and Biomolecular Engineering, North Carolina State University, Raleigh, NC, United States*

1. Introduction

Analysis of single cells or particles is one of the most basic and important techniques in medical diagnosis and healthcare. For clinical purposes, it is essential and preferable to perform quick biological assays in highly heterogeneous systems with large cell populations, such as the blood samples [1]. Optical microscopy, which is by far the most commonly used imaging approach for cellular analysis, has the ability to acquire cell images, with a detailed illustration of the internal/external structure and features of the subunits of cells, including membrane receptors, subcellular organelles, and small intracellular molecules [2]. Nonetheless, most of the microscopic techniques lack the capability of analyzing a considerable number of targeted cells simultaneously due to their relatively low throughput, which greatly restricts their applications in biomedical studies that require high-throughput analysis of biological samples containing a large population of individual cells [3]. Analysis of blood cells, including the cellular evaluation (e.g., blood cell counting) and the molecular profiling (e.g., quantification of biomarkers), is a standard and key method for the determination of a variety of diseases. A system that enables quick acquisition and examination of a large population of blood cells is highly desirable.

With the continuous development of microscopic and biomolecular imaging technology, flow cytometry has been rapidly developed in the past few decades, and has become one of the most powerful techniques for cell analysis and clinical diagnosis throughout the fields of life science [4–6]. Conventional flow cytometer typically allows for high-throughput imaging and characterization of one cell at a time, by hydrodynamic focusing a stream of cells into a fluidic system that is governed by a sheath flow with a flow rate up to 100,000 cells per second (Fig. 5.1) [7,8]. The focused cell stream (typically 10–30 μm in diameter) is then scanned by a focused laser beam ($\sim$20–40 μm width), as it passes perpendicularly through the center of the flow channel ($\sim$50 μm). The subsequent analysis of scanned cells by built-in optical or electronic modules enables the acquisition of useful information on individual cells regarding their size, shape, morphology, granularity, and surface chemistry [4].

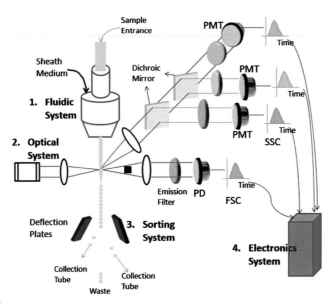

FIGURE 5.1

Configuration of a conventional fluorescence-based flow cytometer, which consists of (A) fluidic system to generate highly focused single-cell stream; (B) optical illumination and detection system, including forward light scatters (FSCs), orthogonal (side) light scatters (SSCs), fluorescence detectors (PDs) and photomultipliers (PMTs); (C) cell sorting and collection system; and (4) electronic control system.

Reproduced with permission from Ref. [8]. Copyright 2010 American Institute of Physics.

Despite the tremendous success achieved, traditional flow cytometry often requires complicated and bulky instrumentation as well as highly trained personnel for operation and data acquisition. As such, flow cytometry is primarily used in research laboratories and less applicable to field tests, especially in resource-limited settings. Some of those obstacles have been promisingly overcome by employing state-of-the-art microfluidic techniques: cost-effective and easy-to-use microfluidic devices provide alternative strategies for effective liquid handling and in situ detection of analytes in tiny volumes of complicated aqueous samples [9,10]. Using various modern micro- or nanofabrication methods such as photolithography, electron beam (e-beam) lithography, and thin-film deposition or etching, microfluidic devices that can be deployed broadly for clinical testing can also be massively produced.

The widespread handheld electronic communication devices in combination with miniaturized diagnostic assays have created new possibilities for inexpensive point-of-care (POC) diagnostics and healthcare delivery [11,12]. Portable bio-analytical systems have been developed by integrating optical, electrical, or mechanical components with mobile consumer devices that enable on-site, rapid, reproducible, and sensitive imaging and sensing of biologically relevant targets of interest [13]. Among various options for miniature imaging devices, smartphones with built-in and compact cameras have drawn tremendous attention due to their cost-effectiveness, portability, and significantly improved optical performance. Smartphone-based devices have been extensively employed as portable readers for uses in POC diagnostics [14—20]. Smartphone detectors could be used by untrained personnel to perform routine bioanalytical measurements, which offer great potentials for improving disease diagnosis in low-resource settings [21,22].

Flow cytometry has undergone drastic transitions from a routine clinical technique to a multipurpose diagnostic system when interfacing microfluidics and smartphone-based devices. This chapter will provide a historical perspective on the development of conventional flow cytometry and a brief summary of recent progress on smartphone-based detectors integrated with microfluidic flow modules as the new generation of miniaturized, portable flow cytometers. Specific emphasis will be placed on recent advances in the microfluidic focusing systems and on-chip flow or imaging cytometry, as well as their relevant clinical applications. The chapter will conclude with regulatoy considerations regarding the use of mobile health devices in clinical settings and a brief perspective on future development of mobile flow cytometry.

2. Traditional flow cytometry

Flow cytometric approaches for the identification and characterization of cellular signals, over time, have evolved into various fundamental biomedical applications within the clinical laboratory. Flow cytometry has a relatively short history dating back to ~70 years ago. With the discovery of Coulter Principle in the early 1950s and the invention of cell sorters by Fulwyler in 1965, optical detectors soon became commercially available. The first fluorescence-based flow cytometer came out in the late 1960s [23]. Owing to the great efforts on standardization and optimization of instrumentation, flow cytometry has achieved rapid development in the next few decades. Nowadays, it has been routinely used as a diagnostic apparatus for a wide range of diseases, such as malaria, leukemia [24], colon cancer, and autoimmune diseases (e.g., human immunodeficiency virus, or HIV) [25].

Flow cytometry is a useful technique for the analysis of multiple physical and chemical characteristics of a large population of cells or particles in real time. In a typical flow cytometer, the cellular analysis is performed by passing a single-file stream of cells through a focused excitation light beam at a rate from hundreds of cells to over 10^6 cells per second (Fig. 5.1) [8]. Through the analysis of fluorescence emissions or scattered light resulting from individual cells, information regarding the size, shape, type, and content of cells can be directly or indirectly obtained. When equipped with cell sorters, the flow cytometer can selectively isolate cells or particles with specified optical properties at constant rates. A conventional cytometer system generally explores dielectric (impedance-based) [26–32], forward angle scattered or side scattered (scattering-based) [33–36] properties of cells, or relies on fluorescent intensities of attached fluorescent dyes or fluorescent dye-tagged monoclonal antibodies [34,35,37,38] to characterize the surfaces or internal components of individual cells.

Although conventional flow cytometry can analyze up to 100,000 cells per second, it is subject to two major drawbacks: first, it commonly lacks the ability to give sufficient multidimensional, intracellular spatial information, as usually only a single feature is measured per scan, especially for fluorescence-based flow cytometry; second, it is largely a laboratory-based technique due to its bulk instrumentation, and therefore normally limited to lab uses [39,40]. For the former, the current trend is to expand the number of parameters that can be simultaneously measured by building instruments with multiplexed optical modules in combination with highly sensitive and chemically specific probes. For example, by replacing fluorescent probes with mass isotope probes, a recently developed mass cytometry can measure approximately 40 markers simultaneously using antibodies tagged with rare earth metals [41]. This platform significantly increases the number of parameters measured beyond what is currently available with conventional flow cytometry and has intensively employed machine learning techniques for the analysis of the multiplexed assays. The latter, on the other side, requires miniaturization of the sampling or imaging devices for the development of the next generation of versatile, compact and portable flow cytometers. Recent attempts toward this direction are highlighted later.

3. Microfluidic flow cytometry

Microfluidic flow cytometry has become an emerging tool for rapid characterization of particles and cells. With recent technical advances, various microfluidic devices that use a set of microchannels etched or molded into polymeric or glass materials have been applied to flow cytometry and extensively used in academic or clinical research. Compared to the traditional benchtop counterparts that suffer from several limitations, cytometers made of less complicated microfluidic channels have a remarkable advantage of reduced sample volume (in the range of nL to fL scale), and thereby much lower operational costs. The introduction of microfluidic flow parts also enables effective miniaturization of conventional set-up and endows cytometric devices with ideal portability and simplicity. The spatial resolution of microfluidic flow cytometers has also been significantly improved, which permits the imaging of not only whole cells, but also a variety of biologically important macromolecules or small molecules, including DNA, RNA, oligonucleotides, proteins, chromosomes and hormones. Moreover, microfluidic cytometers perform excellently in studying the subunits of cells, including membrane [42], cytoplasm [43] and nucleus [44], and therefore has played pivotal roles in various biological and medical applications.

The optimization of microfluidic focusing methods (i.e., methods for accurate positioning of cells in a liquid jet either hydrodynamically or acoustically) is critical to the improved performance of microfluidics-based flow cytometers. Mainstream hydrodynamic focusing systems are based on two-dimensional [9,45], three-dimensional [46], dielectrophoresis/electrokinetic [45], acoustic [47] or inertial flow [48] techniques. Microfluidic flow focusing methods not only confine the particles to a single-file flow in a manner similar to the conventional flow cytometers (by exerting a surrounding sheath stream), but also ensure a uniform particle velocity through the prevention of potential collisions between the flowed cells. It greatly eliminates the potential abnormal parabolic flow profiling that would otherwise exist and diminish the accuracy and reproducibility of the detection results [49]. Among the various focusing systems, 2D microfluidic focusing is most widely employed, which possesses the advantages of simple fabrication, facile driving, low power demand, and straightforward integration. A typical example of a 2D hydrodynamic focusing microfluidic system is shown in Fig. 5.2. However, 2D focusing systems inherently have low detection sensitivity due to the factor that the sample can only be focused on the horizontal plane. To that end, many other effective focusing means, such as 3D, dielectrophoresis, acoustic, and inertial flow methods have been developed to provide better spatial control of cell positions and reliability of detected signals.

Efficient counting of cells is another critical factor in evaluating the quality of a flow cytometer for various biomedical assays. Preseparation of target cells from highly heterogeneous biological samples is necessary. In principle, it is preferable

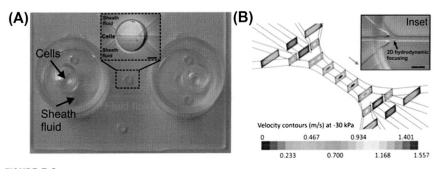

FIGURE 5.2

A microfluidic cytometer based on the 2D hydrodynamic focusing method. (A) Top-down view of the on-chip microfluidic flow cytometer. Inset: microphotograph showing the magnified section of the chip (marked with blue dotted lines). (B) Contours of fluid flow velocity obtained by computational fluid dynamic simulations. Inset: phase-contrast microphotograph depicting the 2D hydrodynamic focusing inside the microfluidic cytometer.

Reproduced with permission from Ref. [9]. Copyright 2013 Elsevier.

to achieve as high throughput as possible. However, insufficient spatial resolution may arise in exchange for the desired throughput. Current image analysis systems typically reach a particle counting rate of 10^3-10^5 cells/s to guarantee the optimal quality of cellular imaging profiles.

Although the field of microfluidic flow cytometry has experienced impressive progress and accomplishments over the past decade, new technology is always in pressing need to further reduce the size, cost, and complexity of cytometric systems. Meanwhile, the analytical sensitivity or specificity of microfluidic flow cytometry needs to be enhanced by bringing in more effective flow control machinery (e.g., inertial focusing with sheathless microflow systems) or by developing more chemically responsive and optically active labels.

4. On-chip imaging cytometry

4.1 Lens-based imaging system

Obtaining spatial information of cells at high throughput is always challenging. On-chip imaging cytometry addresses this issue by integrating high-throughput microfluidics with fast microscopy methods at the single-cell resolution to provide detailed temporal and spatial information simultaneously [34,42,46,50]. The recent advances in optics and digital imaging methods allows for the acquisition of the profiles of thousands of individual cells within minutes. Similar to many other cytometric techniques, on-chip imaging cytometry is suitable for analyzing a large number of nonadherent cells, which is a key feature for many clinical applications such as the analysis of blood samples.

Many sophisticated designs of lens-based on-chip imaging platforms and their potential applications in biomedical detection have been reported. Rasooly and coworkers developed a series of webcam-based, cell steak imaging cytometers for wide-field detection of rare blood cells [51–53]. Gorthi and coworkers have demonstrated several prototype field-implementable imaging devices, for example, a multiple field-of-view imaging flow cytometer supported by a diffractive lens for blood or leukemia cell counting [54] and a microfluidic microscopy system using objective and tube lens for the quantification of yeast cell viability [55] or cancer screening [56]. Han et al. reported a spatial-temporal transformation technique used in a traditional flow cytometer to acquire fluorescence and backscattering images of cells traveling in fluid (Fig. 5.3) [44]. Zhu et al. reported a cellphone-based optofluidic imaging cytometer with the use of an inexpensive cellphone-based fluorescence microscope (Fig. 5.4A). In this study, fluorescently labeled cells were continuously sampled through a microfluidic chip and transported to the imaging field of view (FOV) using a syringe pump. The cells were then excited by low-cost LEDs through waveguide coupling. The device exhibited excellent imaging performance that is comparable to the commercially available hematology analyzer, with a spatial resolution of ~ 2 µm [38]. Lens-assisted imaging flow cytometry has also proven a

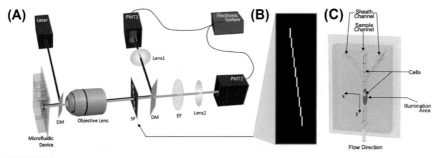

FIGURE 5.3

Schematic illustration of an on-chip imaging cytometer system based on a spatial filter. (A) structure of the on-chip imaging cytometer system. (B) magnified spatial filter design that contains 10 parallel slits (0.1× 1 mm). (C) details of a microfluidic device, with suspended cells focused by a sheath flow to travel in the microfluidic channels.

Reproduced with permission from Ref. [44]. Copyright 2015 Nature Publishing Group.

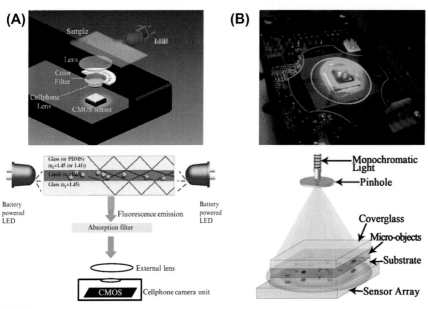

FIGURE 5.4

Lens-based versus lensfree portable flow cytometry. (A) a smartphone-based imaging flow cytometer based on an opto-microfluidic setup. (B) schematic illustration of a lensfree 2D Holographic-LUCAS imaging platform using a CMOS sensor array.

Reprinted with permission from Refs. [38,64], respectively. Copyright 2011 American Chemical Society and 2009 Royal Society of Chemistry.

valuable system especially for rapid and sensitive screening of water quality, for example, for monitoring waterborne pathogen concentrations [57].

4.2 Lensfree imaging system

The rapid development of image sensors has facilitated the great advances in miniature imaging technologies, such as the lensfree on-chip cytometry [40,58]. Lensfree on-chip imaging modality makes it feasible to analyze large volumes of specimen within a short time frame due to its extraordinary large FOV. Because of its compact and light-weight imaging architecture, lensfree imaging has shown great promise for multi-parameter single-cell analysis in the limited resource settings [59].

Lensfree imaging modality typically places the sample plane to the close proximity of the image sensor without using any optical lenses. The samples were illuminated by partially coherent light and their shadows were protected to the image sensor, which contained rich intensity and phase information of the objects. Digital processing of the generated diffraction patterns (shadows) allows for the reconstruction of the images of objects. These compact on-chip platforms offer comparable image resolution to those of conventional lens-based microscope, and are also capable of three-dimensional representations of cells or microparticles through computational image processing [60–63].

Due to the extended field of view of lensfree imaging, thousands of diffraction or shadow patterns of cells or microobjects can be characterized in a single image frame. Lensless imaging techniques have achieved a remarkable breakthrough in clinical diagnostic applications. For example, Ozcan et al. have developed holographic shadow imaging cytometry, or "Holographic-LUCAS," for high-throughput single-cell imaging and counting (Fig. 5.4B) [64–66]. In this technique, on-chip holographic diffraction patterns were recorded using a high resolution image sensor array, and then processed by using a custom developed algorithm for rapid image reconstruction. This system has proved useful for analyzing biological solutions ranging from blood cells, fungi, viruses, to microparticles without the use of any expensive lenses or microscopic pieces. The system was first applied to identify and differentiate image patterns of different microobjects with similar sizes (i.e., microbeads or yeast cells, with $\sim 10\ \mu m$ in diameter) [64]. Later, The performance of lensfree on-chip imaging has been significantly improved to achieve subcellular resolution for the analysis of microorganisms [60]. When combining with image reconstruction algorithms, various applications for cell counting have been demonstrated [62,67–69].

5. Smartphone-based cytometry

Cost-effective diagnostic tools are highly demanded for accurate determination of health conditions, especially in poorly resourced regions. These techniques must be reasonably simple, sufficiently multifunctional, and readily accessible in these settings. Additionally, those devices must be suitable for on-site and point-of-care measurement to serve a large population of people. In the past decade, a variety of handheld optoelectronic devices have been demonstrated for biomedical sensing and diagnostic applications in complicated biospecimens [70–73].

With the explosive growth of mobile technologies, more and more consumer electronics and wireless devices (e.g., smartphones, tablets, flatbed scanners, and webcams) are now equipped with high-performance optics and image sensors, making them attractive as cost-effective biomedical imaging/sensing instrument. Among various choices for such portable devices, smartphones with built-in, miniaturized cameras [14–18] have drawn tremendous attention due to their relatively cheap cost, high portability and high availability. Smartphones have become increasingly popular all over the world, with an estimated total sale of over 1.5 billion devices worldwide in 2017 [74]. Smartphones provide ubiquitous opportunities for easy translation of laboratory biomedical diagnostics into mobile telemedicine tests, which offers new opportunities for the medical practice in both developed and developing countries. Smartphones also carry advanced processing units, which allow on-site rapid data analysis for various mobile health applications.

With the continuous update of camera and CMOS imaging sensors, smartphone-based imaging tools are capable of competing with benchtop microscopes in image contrast and sensitivity. The spatial resolution of smartphone microscopy has been significantly improved to reach 1–3 μm. In this way, utilizing smartphone camera as a sensor to conduct single-cell analysis on a cellphone becomes a practical idea. Compared with flow cytometry, imaging cytometry is more easily integrated with smartphone readers due to the intrinsic large FOV of smartphone devices. These new approaches provide promising alternative opportunities in POC blood cell analysis.

6. Diagnostic applications of smartphone-based cytometry
6.1 Blood cell imaging and counting

Clinically known as complete blood count (CBC), counting blood cells is a fundamental test in blood sample analysis. Many diseases can be diagnosed by the information given by blood cell counts. For example, an abnormal count of white blood cells (WBC) can indicate infection, inflammation, or leukemia. Low red blood cell (RBC) count is often associated with anemia. Blood cells can be manually counted by using a cell counting chamber, namely hemocytometer, under the assistance of an optical microscope. As the emergence of early flow cytometers in the 1950s, counting and analyzing blood cell has been a major application of flow cytometer. Today, automatic hematology analyzer can give RBC counts, platelet counts, 5-part differential WBC counts including lymphocytes, monocytes, neutrophils, eosinophils, and basophils, as well as numbers of abnormal cells simultaneously [75]. Although these cutting-edge instruments are widely used in laboratories and clinics, they are also expensive and bulky, limiting their applications in resource-limited or field settings.

The earlier work in blood cell counting via smartphone-based cytometry was reported by Zhu et al. where a portable optofluidic imaging cytometry platform was developed by combining a microfluidic chamber with smartphone-based fluorescence imaging. The device consists of an LED-based imaging unit, a microfluidic chamber, and a syringe pump for flow control. The WBCs in the fresh blood sample were fluorescently labeled with nucleic acid stain and diluted before delivered into the microfluidic chamber by a syringe pump. By capturing and analyzing the videos of cell movement in the chamber, the concentration of labeled WBC was calculated. The result showed good agreement with a commercial hematology analyzer [38]. In follow-up work, a smartphone-based whole blood analyzer with three switchable testing units was developed (Fig. 5.5A). In this work, removable add-on attachments were connected to the base attachment to perform portable WBC counting, RBC counting, and hemoglobin (Hb) measurement on the same device by using fluorescence imaging, brightfield imaging, and colorimetric detection, respectively. The

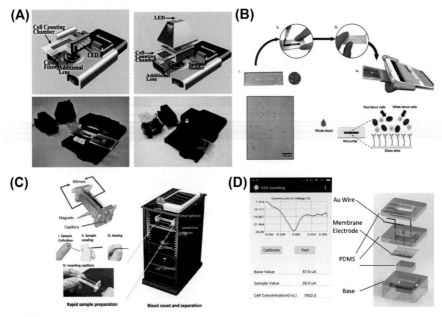

FIGURE 5.5

Smartphone-based cytometry for blood cell imaging and counting. (A) smartphone-supported imaging cytometry platform with switchable add-on components for whole blood analysis [76]. (B) capturing and imaging of CD4 cells on a smartphone platform for the evaluation of HIV infection [78]. (C) blood cell separation and concentration quantification on a smartphone-based magnetic levitation [83]. (D) electrochemical sensor for blood cell counting controlled by a smartphone [93].

Reproduced with permission from the references mentioned earlier. Copyright 2013 Royal Society of Chemistry, 2017 Royal Society of Chemistry, 2016 Wiley, and 2017 Elsevier.

WBC counting attachment consists of eight LEDs as excitation light source and a low-cost color filter as emission filter. WBCs were fluorescently labeled and then imaged in a cell counting chamber. On the other side, the RBC counting was based on brightfield imaging, where the attachment contained a single white LED and no color filters. The lens of this attachment also had a smaller focal length (f = 4 mm) for better resolution and a smaller FOV, as RBCs have a much higher density but smaller size in the blood than WBCs. The principle of Hb measurement was similar to pulse oximetry, although a different wavelength was used. Instead of the red and infrared wavelengths in traditional pulse oximetry, the absorbance at 430 nm was used to quantify Hb concentration through a colorimetric assay. This was realized by capturing an image of a lysed blood sample in front of a blue LED. Transmitted light intensity was proportional to the pixel brightness and therefore the light absorbance as well as the Hb concentration can be calculated [76].

In HIV infection testing and management, CD4+ T lymphocyte (CD4 cell) count remains to be an important diagnostic test. Depletion of CD4 cell was found to be associated with HIV infection. Fluctuation of CD4 cell count reflects the progression of infection and efficacy of antivirus treatment. Thus, CD4 cell count provides vital information in AIDS diagnosis, therapeutics and disease surveillance [77]. Currently, CD4 counts are obtained by conventional flow cytometers, which limits their applications in the developing world due to the lack of medical infrastructure. With the development of smartphone imaging technology, automated CD4 counting platform has recently been enabled on a smartphone (Fig. 5.5D). In this work, as the blood sample flowed through an anti-CD4 antibodies functionalized microfluidic channel, CD4 cells were captured on the surface of the chamber. After rinsing with buffer, the chamber was then placed into the smartphone microscope for imaging and quantification of captured cells. Both preisolated CD4 cells and whole blood sample were evaluated on the smartphone platform, and were compared with results from hemocytometer or flow cytometer. The system showed similar sensitivity with the flow cytometer in the test, which demonstrated great potential as a cost-effective device for monitoring and management of HIV infection in a low-resource setting [78].

Blood cell morphologies can also be diagnostic markers for a number of blood-related diseases such as sickle cell disease. In early studies, smartphone microscopy has been demonstrated for identification of sickle cells in a blood smear [79]. Later on, effective cell separation methods have been developed to facilitate sickle cell isolation and counting. For example, magnetic levitation has been utilized for sickle cell separation and detection on a handheld diagnostic platform [80]. Knowlton et al. designed a smartphone-based device for the magnetic levitation and identification of sickle cells. This device contains a light source, a lens system, a capillary tube and two permanent magnets for levitation. Sickle cells were found to stay at different height compared with normal cells during the levitation because of the density difference [81]. Magnetic levitation-based smartphone cytometry has also been applied in the separation of many other blood cells, including RBCs, WBCs, and platelets. For example, WBCs were successfully separated from RBCs with this technology, and WBC counts can further be obtained from the brightfield images taken by the smartphone [82]. Recently, a magnetic levitation-based smartphone cell quantification platform, named i-LEV, was developed for rapid blood cell analysis (Fig. 5.5C). In this work, the cell count was correlated with the width of cell suspension bands through a linear relationship [83].

Many blood-borne parasitic diseases are threatening millions of people in Africa. Malaria is one of the most deadly and infectious blood-borne parasitic diseases. Rapid detection and diagnosis are especially meaningful to control the spread of the disease. Automation of malaria diagnosis has become a trend with a number of image processing methods being adapted to count the number of blood cell and parasites [84]. Based on these methods, malaria diagnosis has been made available on mobile phone devices. As one of the earliest studies in this field, Breslauer et al. reported mobile phone imaging of Giemsa-stained malaria-infected blood sample, which can be used in malaria screening [79]. A mobile phone-based polarized microscope was also recently developed by Pirnstill et al. to detect hemozoin, a malaria-related pigment for malaria diagnosis. Hemozoin crystal is birefringent and can be observed under polarized light conditions. The results from the smartphone were compared with the result from the polarized microscope and the positions of hemozoin crystals in blood smear samples were well correlated [85]. Rosado et al. reported an automated detection method of malaria by observing WBCs with a smartphone microscope for the detection of trophozoites, a growing stage of the malarial parasite. The blood smear was magnified by an optical lens tube and images were then analyzed by a smartphone application. Image segmentation was conducted automatically based on geometry, color and texture of cellular structure to extract WBC nuclei and trophozoites [86].

Besides lens-based smartphone imaging techniques, lensfree imaging has also been applied in smartphones for blood cell counting and analysis. Lensfree imaging on the mobile phone was first reported in 2010, in which holographic pattern created by an aperture was projected directly on the image sensor and was used for image reconstruction [87]. In another example, sun light was used as light source for a lensfree smartphone on-chip microscope. The image was reconstructed from multiple images captured at different angles of illumination [88]. More recently, a smartphone-based lensfree analysis platform called Digital Diffraction Diagnosis (or D3 assay) was reported for single-cell imaging and cancer biomarker detection. In this assay, diffraction patterns of immunolabeled microbeads and cells were captured and processed to retrieve transmittance and phase contrast and reconstruct single-cell image. By analyzing the amount of microbeads bound to the target cells, different kinds of tumor cells could be identified [89]. In the case mentioned earlier, blood cells, plankton cells, and tumor cells could be successfully reconstructed from holographic or diffraction patterns captured by a smartphone, demonstrating its potential for portable single-cell imaging and counting in the field settings. When applied for cytometry measurement, lensfree cytometer provides an even larger FOV while keeping a much simpler instrumental setup than lens-based smartphone cytometry. For example, Roy et al. developed a platform for label-free blood cell counting on the smartphone. Based on the difference in shadow patterns, RBCs and WBCs were identified and counted from whole blood samples and the results were validated by a commercial blood analyzer. Moreover, the major subtypes of WBCs—neutrophils, lymphocytes, and monocytes were also differentiated and counted respectively from lysed blood sample [90].

Smartphone can also serve as an interfacing device to display and analyze data acquired from other biosensors to construct a portable cytometer for personalized health monitoring and diagnostics. For example, Talukder et al. reported a smartphone-mediated impedance cytometer for RBC counting in polydimethylsiloxane (PDMS) microfluidic channels [91]. A similar device based on a flexible circuit board with online readout function was developed by Furniturewalla et al. [92]. Recently, a paper-based electrochemical biosensor was reported by Wang et al. for WBC counting (Fig. 5.5D) [93]. In this cyclic voltammetry-based sensor, an increase in the concentration of cells leads to a decrease of the reduction peak due to the blockage of diffusion caused by the trapped cell in the paper matrix. A linear trend between the cell counts from the portable system and the manual counting results was observed, proving a good accuracy of this sensing platform.

6.2 Cancer cell imaging and counting

Knowlton et al. demonstrated a smartphone-based imaging and magnetophoretic cytometry device for density-based cell sorting and cancer cell detection (Fig. 5.6). Magnetic levitation was used again for counting and separating cells in a paramagnetic medium under an external magnetic field. Fluorescence images of suspended breast cancer cells were taken and cell counts were obtained subsequently. The counting results were compared with a commercial hemocytometer

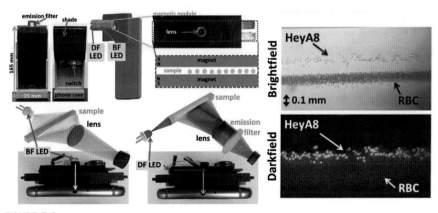

FIGURE 5.6

Smartphone cytometry using magnetic levitation and fluorescence microscopy for counting and separation of tumor cells.

Reproduced with permission from Ref. [37]. Copyright 2017 Royal Society of Chemistry.

and a linear relationship was found. In addition, ovarian cancer cells were mixed with red blood cells, and they were successfully separated with magnetic levitation because of the difference in density, which was verified with both smartphone bright-field and fluorescence imaging [37]. Similarly, using magnetic levitation, separation of blood cells and cancer cells was achieved on a cost-effective smartphone accessory fabricated by 3D printing. The device was tested to be effective in separating blood cells with breast cancer cells, lung cancer cells, ovarian cancer cells and prostate cancer cells [94].

6.3 Pathogen, parasite, and microswimmer imaging and counting

Rapid detection and quantification of parasites from human or environmental samples is important in parasitic disease control, screening and prevention in the developing countries. Increasing examples of using smartphones for pathogen, parasite, and microswimmer detection and counting have been demonstrated. For example, blood-borne filarial parasites *Loa loa* has been detected and quantified on a smartphone video microscope recently (Fig. 5.7A). A video of blood sample flowing in a microfluidic chamber was recorded at first. Difference images between each video frame and the time-averaged frame were then calculated. If parasites were present,

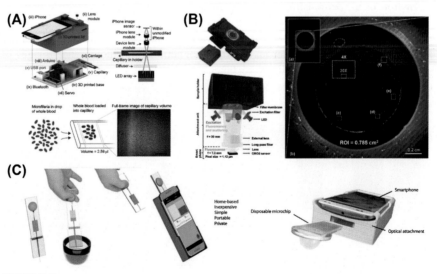

FIGURE 5.7

Smartphone cytometry for pathogen, parasite, and microswimmer counting. (A) blood-borne filarial parasite counting on a smartphone video microscope [95]. (B) detection and quantification of *Giardia lamblia* cysts by a smartphone based fluorescence microscope [96]. (C) POC semen analysis platform on a smartphone [100].

Reproduced with permission from the references mentioned earlier. Copyright 2015 American Association for the Advancement of Science and 2017 American Association for the Advancement of Science.

signals would show on these difference images which could be identified automatically using an appropriate algorithm. Sample flow and illumination were controlled using a microcontroller. The results could be displayed directly in an iOS app [95]. Smartphone-based detection and quantification of waterborne parasite *Giardia lamblia* was reported by Koydemir et al. (Fig. 5.7B). Cysts of the parasite were fluorescently labeled and imaged under a smartphone fluorescence microscope. The accurate count of cysts was given from cloud analysis based on a machine learning algorithm [96]. Besides parasites, detection and quantification of other pathogens such as bacteria and viruses have also been reported on smartphone platforms. Wu et al. reported the detection of *E. coli* with a limit of detection of 10 bacteria/mL. Immunolabeled microbeads were added to the sample spiked with *E. coli* forming aggregated clusters. Scattering light from the cluster can be detected by the smartphone image sensor as signal, which is proportional to bacterial counts [97]. Moreover, mobile phone-based imaging and counting single virus particle was made possible recently. Fluorescent labeled human cytomegalovirus particles were imaged on a smartphone with single virus sensitivity confirmed by SEM images. As a step toward nanosensing and imaging, this technique show potential in high-sensitivity virus detection and quantification in the field settings [98].

On the other side, applications of smartphone-based cell counting techniques for semen analysis by spermatozoa counting have been put forward recently. Kobori et al. reported a simple smartphone microscope used in the semen analysis. A ball-lens was added in front of the smartphone camera for magnification. The number and motility of spermatozoa were counted and measured by analyzing the video of spermatozoa in the semen sample. The system was tested on three different phone models and similar results were given between them [99]. As a step forward, Kanakasabapathy et al. developed an automated smartphone-based semen analysis assay (Fig. 5.7C). Brightfield videos of spermatozoa were taken under the magnification of two aspheric lenses and illumination of a LED light. A low-power sample processing device was demonstrated, where semen samples were drawn into the counting chamber by manually generated vacuum in the microchip. The chip was then inserted into the smartphone accessory for video capturing. The concentration and motility of sperm can be calculated from the video by an Android application-based detection and tracking algorithm. Using this system, untrained users can then test semen quality conveniently at home [100].

7. Regulatory issues, medical requirements, and constraints

The thriving development of novel mobile health technologies has also posed harsh legal challenges to the existing framework of federal regulation and societal obligation [101,102]. Smartphone-based cytometry is facing a similar dilemma that the commercialization and marketing of healthcare devices is sometimes delayed by the product regulation related to performance reliability and data privacy. Government agencies such as the U.S. Food and Drug Administration (FDA) and National Institute of Health have set core standards to which new biomedical applications or devices should strictly adhere. With the increasingly growing use and extensive availability of mobile health devices and applications, both technical developers and healthcare providers (i.e., hospitals and other health systems) are looking for legal guidance to ensure that they comply with regulations and avoid unexpected legal issues [103].

In a sense, any new mobile medical techniques could bring in potential risks and ethical concerns to the public health that would inevitably touch upon traditional areas of healthcare law, including medical malpractice, product copyright, and personal privacy [104]. For example, a newly designed smartphone imaging and sensing device operated in a similar principle to an existing commercial flow cytometer may need to avoid infringement of existing patents when it comes to commercialization. As another example, the use of mobile apps that collect and analyze data from individual patients need to be properly regulated, to prevent the invasion of personal privacy or potential data breach. Scientists and developers therefore need to pay special attention to potential ethical issues involved in the use of various types of mobile health equipment, including smartphone-based cytometers.

To address those abovementioned lawful or ethical concerns, technical developers and healthcare providers are expected to get more involved in the development of rules or protocols that can guide consumers for the best use of those mobile techniques. Government agencies on the other side should keep institutional policies up to date to enhance the regulation through maximizing compliance with medical products and minimizing their risks. A typical example of federal regulations on mobile healthcare use can be found in one of the recent FDA documentation [104]. In this report, it clearly stipulates that private developers of smartphone apps should reduce the liability risks by designing accurate, reliable apps or programs to protect customers' data.

8. Conclusions

With the arrival of the big-data era, scientists from all research fields have to reasonably tackle the challenge with massive volumes of data. Recent advances in microfluidic flow cytometric devices allow continuous high-throughput acquisition of single-cell information from highly complicated biological matrices. Microfluidics provides conventional cytometry with tremendous advantages of significantly reduced instrumental size, cost, sample volume, as well as the integration of functional components for system control. The broad popularity of mobile communication devices, such as smartphones, greatly permits the assembly of miniaturized flow cytometers for rapid, accurate, and inexpensive clinical and biomedical analyses. Various lens-based or lensfree imaging technologies have been implemented on smartphones, making it possible for the generation of multiparametric profiles of a large cell population for both disease detection and health condition monitoring at the POC. Future development of flow cytometry running on mobile phones will continue to focus on improving multimodality (i.e., switchable modes between transmission, scattering, and fluorescence) and compatibility with real biomedical applications including cell counting, sorting, and characterization of particular disease models. In addition, new methods are required for efficient analysis and utilization of this huge amount of data. To sum up, the new trend of developing compact and low-cost mobile flow cytometry will promise to facilitate the advance of fundamental clinical research and impact directly on the improvement of the quality of life globally.

References

[1] Adan A, Alizada G, Kiraz Y, Baran Y, Nalbant A. Flow cytometry: basic principles and applications. Crit Rev Biotechnol 2017;37(2):163—76.

[2] Heintzmann R, Huser T. Super-resolution structured illumination microscopy. Chem Rev 2017;117(23):13890—908.

[3] Goda K, Ayazi A, Gossett DR, Sadasivam J, Lonappan CK, Sollier E, Fard AM, Hur SC, Adam J, Murray C, Wang C, Brackbill N, Di Carlo D, Jalali B. High-throughput single-microparticle imaging flow analyzer. Proc Natl Acad Sci 2012; 109(29):11630—5.

[4] Brown M, Wittwer C. Flow cytometry: principles and clinical applications in hematology. Clin Chem 2000;46(8):1221—9.

[5] Davidson B, Dong HP, Berner A, Risberg B. The diagnostic and research applications of flow cytometry in cytopathology. Diagn Cytopathol 2012;40(6):525—35.

[6] Léonard L, Bouarab Chibane L, Ouled Bouhedda B, Degraeve P, Oulahal N. Recent advances on multi-parameter flow cytometry to characterize antimicrobial treatments. Front Microbiol 2016;7:1225.

[7] Henel G, Schmitz JL. Basic theory and clinical applications of flow cytometry. Lab Med 2007;38(7):428—36.

[8] Cho SH, Godin JM, Chen C-H, Qiao W, Lee H, Lo Y-H. Review Article: recent advancements in optofluidic flow cytometer. Biomicrofluidics 2010;4(4):43001.

[9] Skommer J, Akagi J, Takeda K, Fujimura Y, Khoshmanesh K, Wlodkowic D. Multiparameter Lab-on-a-Chip flow cytometry of the cellcycle. Biosens Bioelectron 2013;42:586—91.

[10] Yamada K, Shibata H, Suzuki K, Citterio D. Toward practical application of paper-based microfluidics for medical diagnostics: state-of-the-art and challenges. Lab Chip 2017;17(7):1206—49.

[11] Vashist SK. Point-of-Care diagnostics: recent advances and trends. Biosensors 2017; 7(4):62.

[12] Myers FB, Lee LP. Innovations in optical microfluidic technologies for point-of-care diagnostics. Lab Chip 2008;8(12):2015—31.

[13] Vashist SK, Schneider EM, Luong JHT. Commercial smartphone-based devices and smart applications for personalized healthcare monitoring and management. Diagnostics 2014;4(3):104—28.

[14] Vashist SK, van Oordt T, Schneider EM, Zengerle R, von Stetten F, Luong JHT. A smartphone-based colorimetric reader for bioanalytical applications using the screen-based bottom illumination provided by gadgets. Biosens Bioelectron 2015; 67:248—55.

[15] Long KD, Woodburn EV, Le HM, Shah UK, Lumetta SS, Cunningham BT. Multimode smartphone biosensing: the transmission, reflection, and intensity spectral (TRI)-analyzer. Lab Chip 2017;17(19):3246—57.

[16] Roda A, Michelini E, Zangheri M, Di Fusco M, Calabria D, Simoni P. Smartphone-based biosensors: a critical review and perspectives. Trac Trends Anal Chem 2016; 79:317—25.

[17] Zhang D, Liu Q. Biosensors and bioelectronics on smartphone for portable biochemical detection. Biosens Bioelectron 2016;75:273—84.

[18] Lopez-Ruiz N, Curto VF, Erenas MM, Benito-Lopez F, Diamond D, Palma AJ, Capitan-Vallvey LF. Smartphone-based simultaneous pH and nitrite colorimetric determination for paper microfluidic devices. Anal Chem 2014;86(19):9554—62.

[19] Kaarj K, Akarapipad P, Yoon J-Y. Simpler, faster, and sensitive zika virus assay using smartphone detection of loop-mediated isothermal amplification on paper microfluidic chips. Sci Rep 2018;8(1):12438.

[20] Cho S, Park TS, Nahapetian TG, Yoon J-Y. Smartphone-based, sensitive µPAD detection of urinary tract infection and gonorrhea. Biosens Bioelectron 2015;74:601—11.

[21] Bates M, Zumla A. Rapid infectious diseases diagnostics using Smartphones. Ann Transl Med 2015;3(15):215.

[22] Kanchi S, Sabela MI, Mdluli PS, Inamuddin, Bisetty K. Smartphone based bioanalytical and diagnosis applications: a review. Biosens Bioelectron 2018;102: 136—49.

[23] Fulwyler MJ. Electronic separation of biological cells by volume. Science 1965; 150(3698):910—1.

[24] Liang S, Slattery MJ, Wagner D, Simon SI, Dong C. Hydrodynamic shear rate regulates melanoma-leukocyte aggregation, melanoma adhesion to the endothelium, and subsequent extravasation. Ann Biomed Eng 2008;36(4):661—71.

[25] Greve B, Weidner J, Cassens U, Odaibo G, Olaleye D, Sibrowski W, Reichelt D, Nasdala I, Göhde W. A new affordable flow cytometry based method to measure HIV-1 viral load. Cytometry 2009;75A(3):199—206.

[26] Cheung K, Gawad S, Renaud P. Impedance spectroscopy flow cytometry: on-chip label-free cell differentiation. Cytometry 2005;65A(2):124—32.

[27] Holmes D, Pettigrew D, Reccius CH, Gwyer JD, van Berkel C, Holloway J, Davies DE, Morgan H. Leukocyte analysis and differentiation using high speed microfluidic single cell impedance cytometry. Lab Chip 2009;9(20):2881—9.

[28] Evander M, Ricco AJ, Morser J, Kovacs GTA, Leung LLK, Giovangrandi L. Microfluidic impedance cytometer for platelet analysis. Lab Chip 2013;13(4):722—9.

[29] Sun T, Morgan H. Single-cell microfluidic impedance cytometry: a review. Microfluid Nanofluidics 2010;8(4):423—43.

[30] Hassan U, Ghonge T, Reddy Jr B, Patel M, Rappleye M, Taneja I, Tanna A, Healey R, Manusry N, Price Z, Jensen T, Berger J, Hasnain A, Flaugher E, Liu S, Davis B, Kumar J, White K, Bashir R. A point-of-care microfluidic biochip for quantification of CD64 expression from whole blood for sepsis stratification. Nat Commun 2017; 8:15949.

[31] Watkins NN, Hassan U, Damhorst G, Ni H, Vaid A, Rodriguez W, Bashir R. Micro-fluidic CD4$^+$ and CD8$^+$ T lymphocyte counters for point-of-care HIV diagnostics using whole blood. Sci Transl Med 2013;5(214):214ra170.

[32] Hassan U, Watkins NN, Reddy Jr B, Damhorst G, Bashir R. Microfluidic differential immunocapture biochip for specific leukocyte counting. Nat Protoc 2016;11:714.

[33] Steen HB. Flow cytometer for measurement of the light scattering of viral and other submicroscopic particles. Cytometry 2004;57A(2):94−9.

[34] Watts BR, Zhang Z, Xu C-Q, Cao X, Lin M. Integration of optical components on-chip for scattering and fluorescence detection in an optofluidic device. Biomed Opt Express 2012;3(11):2784−93.

[35] Etcheverry S, Faridi A, Ramachandraiah H, Kumar T, Margulis W, Laurell F, Russom A. High performance micro-flow cytometer based on optical fibres. Sci Rep 2017;7(1):5628.

[36] Zhang W, Tian Y, Hu X, He S, Niu Q, Chen C, Zhu S, Yan X. Light-scattering sizing of single submicron particles by high-sensitivity flow cytometry. Anal Chem 2018; 90(21):12768−75.

[37] Knowlton S, Joshi A, Syrrist P, Coskun AF, Tasoglu S. 3D-printed smartphone-based point of care tool for fluorescence- and magnetophoresis-based cytometry. Lab Chip 2017;17(16):2839−51.

[38] Zhu H, Mavandadi S, Coskun AF, Yaglidere O, Ozcan A. Optofluidic fluorescent imaging cytometry on a cell phone. Anal Chem 2011;83(17):6641−7.

[39] Simonnet C, Groisman A. High-throughput and high-resolution flow cytometry in molded microfluidic devices. Anal Chem 2006;78(16):5653−63.

[40] Han Y, Gu Y, Zhang AC, Lo Y-H. Review: imaging technologies for flow cytometry. Lab Chip 2016;16(24):4639−47.

[41] Nassar AF, Wisnewski AV, Raddassi K. Automation of sample preparation for mass cytometry barcoding in support of clinical research: protocol optimization. Anal Bioanal Chem 2017;409(9):2363−72.

[42] Rane AS, Rutkauskaite J, deMello A, Stavrakis S. High-throughput multi-parametric imaging flow cytometry. Chem 2017;3(4):588−602.

[43] Zhao Y, Wang K, Chen D, Fan B, Xu Y, Ye Y, Wang J, Chen J, Huang C. Development of microfluidic impedance cytometry enabling the quantification of specific membrane capacitance and cytoplasm conductivity from 100,000 single cells. Biosens Bioelectron 2018;111:138−43.

[44] Han Y, Lo Y-H. Imaging cells in flow cytometer using spatial-temporal transformation. Sci Rep 2015;5:13267.

[45] Wu M, Piccini M, Koh C-Y, Lam KS, Singh AK. Single cell MicroRNA analysis using microfluidic flow cytometry. PLoS One 2013;8(1):e55044.

[46] Mao X, Lin S-CS, Dong C, Huang TJ. Single-layer planar on-chip flow cytometer using microfluidic drifting based three-dimensional (3D) hydrodynamic focusing. Lab Chip 2009;9(11):1583−9.

[47] Piyasena ME, Austin Suthanthiraraj PP, Applegate Jr RW, Goumas AM, Woods TA, López GP, Graves SW. Multinode acoustic focusing for parallel flow cytometry. Anal Chem 2012;84(4):1831−9.

[48] Kim J, Lee J, Wu C, Nam S, Di Carlo D, Lee W. Inertial focusing in non-rectangular cross-section microchannels and manipulation of accessible focusing positions. Lab Chip 2016;16(6):992−1001.

[49] Yang R-J, Fu L-M, Hou H-H. Review and perspectives on microfluidic flow cytometers. Sens Actuators B Chem 2018;266:26−45.

[50] Barteneva NS, Fasler-Kan E, Vorobjev IA. Imaging flow cytometry: coping with heterogeneity in biological systems. J Histochem Cytochem 2012;60(10):723−33.

[51] Ossandon M, Balsam J, Bruck HA, Kalpakis K, Rasooly A. A computational streak mode cytometry biosensor for rare cell analysis. Analyst 2017;142(4):641−8.

[52] Balsam J, Bruck HA, Rasooly A. Cell streak imaging cytometry for rare cell detection. Biosens Bioelectron 2015;64:154−60.

[53] Balsam J, Bruck HA, Rasooly A. Webcam-based flow cytometer using wide-field imaging for low cell number detection at high throughput. Analyst 2014;139(17): 4322−9.

[54] Schonbrun E, Gorthi SS, Schaak D. Microfabricated multiple field of view imaging flow cytometry. Lab Chip 2012;12(2):268−73.

[55] Jagannadh VK, Murthy RS, Srinivasan R, Gorthi SS. Automated quantitative cytological analysis using portable microfluidic microscopy. J Biophot 2016;9(6):586−95.

[56] Jagannadh VK, Gopakumar G, Subrahmanyam GRKS, Gorthi SS. Microfluidic microscopy-assisted label-free approach for cancer screening: automated microfluidic cytology for cancer screening. Med Biol Eng Comput 2017;55(5):711−8.

[57] Koydemir HC, Gorocs Z, McLeod E, Tseng D, Ozcan A. Field portable mobile phone based fluorescence microscopy for detection of Giardia lamblia cysts in water samples. SPIE BiOS, SPIE 2015:8.

[58] Shapiro HM, Perlmutter NG. Personal cytometers: slow flow or no flow? Cytometry 2006;69A(7):620−30.

[59] Kim SB, Bae H, Koo K-I, Dokmeci MR, Ozcan A, Khademhosseini A. Lens-free imaging for biological applications. J Lab Autom 2012;17(1):43−9.

[60] Mudanyali O, Tseng D, Oh C, Isikman SO, Sencan I, Bishara W, Oztoprak C, Seo S, Khademhosseini B, Ozcan A. Compact, light-weight and cost-effective microscope based on lensless incoherent holography for telemedicine applications. Lab Chip 2010;10(11):1417−28.

[61] Luo W, Greenbaum A, Zhang Y, Ozcan A. Synthetic aperture-based on-chip microscopy, Light. Sci Appl 2015;4:e261.

[62] Wei Q, McLeod E, Qi H, Wan Z, Sun R, Ozcan A. On-chip cytometry using plasmonic nanoparticle enhanced lensfree holography. Sci Rep 2013;3:1699.

[63] Singh DK, Ahrens CC, Li W, Vanapalli SA. Label-free, high-throughput holographic screening and enumeration of tumor cells in blood. Lab Chip 2017;17(17):2920−32.

[64] Seo S, Su T-W, Tseng DK, Erlinger A, Ozcan A. Lensfree holographic imaging for on-chip cytometry and diagnostics. Lab Chip 2009;9(6):777−87.

[65] Ozcan A, Demirci U. Ultra wide-field lens-free monitoring of cells on-chip. Lab Chip 2008;8(1):98−106.

[66] Su T-W, Seo S, Erlinger A, Ozcan A. High-throughput lensfree imaging and characterization of a heterogeneous cell solution on a chip. Biotechnol Bioeng 2009;102(3): 856−68.

[67] Stybayeva G, Mudanyali O, Seo S, Silangcruz J, Macal M, Ramanculov E, Dandekar S, Erlinger A, Ozcan A, Revzin A. Lensfree holographic imaging of antibody microarrays for high-throughput detection of leukocyte numbers and function. Anal Chem 2010;82(9):3736−44.

[68] Kesavan SV, Momey F, Cioni O, David-Watine B, Dubrulle N, Shorte S, Sulpice E, Freida D, Chalmond B, Dinten JM, Gidrol X, Allier C. High-throughput monitoring of major cell functions by means of lensfree video microscopy. Sci Rep 2014;4:5942.

[69] Seo S, Isikman SO, Sencan I, Mudanyali O, Su T-W, Bishara W, Erlinger A, Ozcan A. High-throughput lens-free blood analysis on a chip. Anal Chem 2010;82(11):4621–7.

[70] Alkasir RSJ, Rossner A, Andreescu S. Portable colorimetric paper-based biosensing device for the assessment of bisphenol a in indoor dust. Environ Sci Technol 2015; 49(16):9889–97.

[71] Nikkhoo N, Cumby N, Gulak PG, Maxwell KL. Rapid bacterial detection via an all-electronic CMOS biosensor. PLoS One 2016;11(9):e0162438.

[72] Estevez MC, Alvarez M, Lechuga LM. Integrated optical devices for lab-on-a-chip biosensing applications. Laser Photonics Rev 2012;6(4):463–87.

[73] Zhang W, Guo S, Pereira Carvalho WS, Jiang Y, Serpe MJ. Portable point-of-care diagnostic devices. Anal Methods 2016;8(44):7847–67.

[74] Number of smartphones sold to end users worldwide from 2007 to 2017 (in million units). 2018. https://www.statista.com/statistics/263437/global-smartphone-sales-to-end-users-since-2007/.

[75] Sullivan E. Hematology analyzer: from workhorse to thoroughbred. Lab Med 2006; 37(5):273–8.

[76] Zhu H, Sencan I, Wong J, Dimitrov S, Tseng D, Nagashima K, Ozcan A. Cost-effective and rapid blood analysis on a cell-phone. Lab Chip 2013;13(7):1282–8.

[77] Ford N, Meintjes G, Pozniak A, Bygrave H, Hill A, Peter T, Davies MA, Grinsztejn B, Calmy A, Kumarasamy N, Phanuphak P, deBeaudrap P, Vitoria M, Doherty M, Stevens W, Siberry GK. The future role of CD4 cell count for monitoring antiretroviral therapy. Lancet Infect Dis 2015;15(2):241–7.

[78] Kanakasabapathy MK, Pandya HJ, Draz MS, Chug MK, Sadasivam M, Kumar S, Etemad B, Yogesh V, Safavieh M, Asghar W. Rapid, label-free CD4 testing using a smartphone compatible device. Lab Chip 2017;17(17):2910–9.

[79] Breslauer DN, Maamari RN, Switz NA, Lam WA, Fletcher DA. Mobile phone based clinical microscopy for global health applications. PLoS One 2009;4(7):e6320.

[80] Yenilmez B, Knowlton S, Yu CH, Heeney MM, Tasoglu S. Label-free sickle cell disease diagnosis using a low-cost, handheld platform. Adv Mater Technol 2016;1(5): 1600100.

[81] Knowlton S, Sencan I, Aytar Y, Khoory J, Heeney M, Ghiran I, Tasoglu S. Sickle cell detection using a smartphone. Sci Rep 2015;5:15022.

[82] Yenilmez B, Knowlton S, Tasoglu S. Self-contained handheld magnetic platform for point of care cytometry in biological samples. Adv Mater Technol 2016;1(9):1600144.

[83] Baday M, Calamak S, Durmus NG, Davis RW, Steinmetz LM, Demirci U. Integrating cell phone imaging with magnetic levitation (i-LEV) for label-free blood analysis at the point-of-living. Small 2016;12(9):1222–9.

[84] Poostchi M, Silamut K, Maude RJ, Jaeger S, Thoma G. Image analysis and machine learning for detecting malaria. Transl Res 2018;194:36–55.

[85] Pirnstill CW, Coté GL. Malaria diagnosis using a mobile phone polarized microscope. Sci Rep 2015;5:13368.

[86] Rosado L, da Costa JMC, Elias D, Cardoso JS. Automated detection of malaria parasites on thick blood smears via mobile devices. Procedia Comput Sci 2016;90:138–44.

[87] Tseng D, Mudanyali O, Oztoprak C, Isikman SO, Sencan I, Yaglidere O, Ozcan A. Lensfree microscopy on a cellphone. Lab Chip 2010;10(14):1787–92.

[88] Lee SA, Yang C. A smartphone-based chip-scale microscope using ambient illumination. Lab Chip 2014;14(16):3056−63.

[89] Im H, Castro CM, Shao H, Liong M, Song J, Pathania D, Fexon L, Min C, Avila-Wallace M, Zurkiya O. Digital diffraction analysis enables low-cost molecular diagnostics on a smartphone. Proc Natl Acad Sci 2015;112(18):5613−8.

[90] Roy M, Jin G, Seo D, Nam M-H, Seo S. A simple and low-cost device performing blood cell counting based on lens-free shadow imaging technique. Sens Actuators B Chem 2014;201:321−8.

[91] Talukder N, Furniturewalla A, Le T, Chan M, Hirday S, Cao X, Xie P, Lin Z, Gholizadeh A, Orbine S. A portable battery powered microfluidic impedance cytometer with smartphone readout: towards personal health monitoring. Biomed Microdevices 2017;19(2):36.

[92] Furniturewalla A, Chan M, Sui J, Ahuja K, Javanmard M. Fully integrated wearable impedance cytometry platform on flexible circuit board with online smartphone readout. Microsyst Nanoeng 2018;4(1):20.

[93] Wang X, Lin G, Cui G, Zhou X, Liu GL. White blood cell counting on smartphone paper electrochemical sensor. Biosens Bioelectron 2017;90:549−57.

[94] Amin R, Knowlton S, Dupont J, Bergholz JS, Joshi A, Hart A, Yenilmez B, Yu CH, Wentworth A, Zhao JJ. 3D-printed smartphone-based device for label-free cell separation. J 3D Print Med 2017;1(3):155−64.

[95] D'ambrosio MV, Bakalar M, Bennuru S, Reber C, Skandarajah A, Nilsson L, Switz N, Kamgno J, Pion S, Boussinesq M. Point-of-care quantification of blood-borne filarial parasites with a mobile phone microscope. Sci Transl Med 2015;7(286):286re4.

[96] Koydemir HC, Gorocs Z, Tseng D, Cortazar B, Feng S, Chan RYL, Burbano J, McLeod E, Ozcan A. Rapid imaging, detection and quantification of Giardia lamblia cysts using mobile-phone based fluorescent microscopy and machine learning. Lab Chip 2015;15(5):1284−93.

[97] Wu T-F, Chen Y-C, Wang W-C, Kucknoor AS, Lin C-J, Lo Y-H, Yao C-W, Lian I. Rapid waterborne pathogen detection with mobile electronics. Sensors 2017;17(6): 1348.

[98] Wei Q, Qi H, Luo W, Tseng D, Ki SJ, Wan Z, Gorocs Zn, Bentolila LA, Wu T-T, Sun R. Fluorescent imaging of single nanoparticles and viruses on a smart phone. ACS Nano 2013;7(10):9147−55.

[99] Kobori Y, Pfanner P, Prins GS, Niederberger C. Novel device for male infertility screening with single-ball lens microscope and smartphone. Fertil Steril 2016; 106(3):574−8.

[100] Kanakasabapathy MK, Sadasivam M, Singh A, Preston C, Thirumalaraju P, Venkataraman M, Bormann CL, Draz MS, Petrozza JC, Shafiee H. An automated smartphone-based diagnostic assay for point-of-care semen analysis. Sci Transl Med 2017;9(382). eaai7863.

[101] Schoenfeld AJ, Sehgal N, Auerbach A. The challenges of mobile health regulation. JAMA Intern Med 2016;176(5):704−5.

[102] Vincent CJ, Niezen G, O'Kane AA, Stawarz K. Can standards and regulations keep up with health technology? JMIR Mhealth Uhealth 2015;3(2):e64.

[103] Parker L, Karliychuk T, Gillies D, Mintzes B, Raven M, Grundy Q. A health app developer's guide to law and policy: a multi-sector policy analysis. BMC Med Inf Decis Mak 2017;17(1):141.

[104] Barton AJ. The regulation of mobile health applications. BMC Med 2012;10:46.

Smartphones for rapid kits

6

Anna Pyayt, PhD

*Associate Professor, Chemical & Biomedical Engineering, University of South Florida, Tampa, FL,
United States*

1. Introduction

The smartphones are a great platform for biomedical diagnostics, as technology advances, such as greatly enhanced processing power, storage capability, and wireless connectivity, turned smartphones into powerful computers integrated with high quality cameras and numerous sensors. Additionally, the price and size of the smartphones have decreased so much that they became available even in the countries with the lowest income. The worldwide mobile phone subscription recently reached nearly seven billion users [1]. Some of the applications of the smartphones in healthcare and biomedical fields include weight management [2], lens-free microscopy [3], hypertension monitoring system [4], label free immunoassays [5], monitoring system for Parkinson's disease patients [6], retinal disease diagnostic device [7], system for monitoring kidney metabolomics [8], flow cytometry [9] and many others [10].

The smartphone-based biomedical testing requires integration with analyte-specific sensors generating signals that can be processed with a visual readout. Because of that, integration with rapid kits is one of the very important areas of mobile health. The types of rapid kits that can be used include colorimetric assay kits, either in a strip (e.g., dipsticks) or a test tube. They usually contain some reagents that react with biological fluids (blood, urine, saliva, etc.), and change color according to the concentration of the measured substance. Often, they are used without any reading equipment, purely with visual assessment, and under those circumstances have low accuracy. Sometimes, many different tests can be simultaneously conducted with the same rapid kit. For example, dipsticks traditionally used for urinalysis can have 10 different pads sensitive to such urine parameters as concentration of glucose, proteins, ketones, and many others, and all of them can be simultaneously analyzed by looking at their color change after 2 min exposure to a urine sample.

However, integration of smartphones with rapid kits should satisfy several additional important constraints less critical for other areas of mobile health. Specifically, these biomedical testing systems must be very accurate, as false positives would require expensive additional testing, while false negatives might result in wrong diagnosis and negative health outcomes for the patient. In addition to that, as the test is often conducted far from a lab or a hospital, and sometimes by personnel without much medical training, they should be easy to perform. The optimal time from the beginning of the test to obtaining the results should be between 5 min and 2 hours, as sometimes it might be critical for important time-sensitive medical decision, or the user might not be willing to wait longer. Finally, the cost is very important, as insurance might be unavailable in resource limited settings or might not be used for confidentiality reason. In such cases, the full cost of the test should be covered by the user, and thus it has to be low, to receive wide adoption.

2. Ways to use smartphones for raid kits

The way how a smartphone can be used for rapid kits analysis depends on a type of biological sample used for the test. For example, sperm testing system can determine sperm concentration, motility, and linear and curvilinear velocities by performing an on-phone image analysis using a semen sample loaded into a low-cost disposable microfluidic device [11]. Alternatively, there are different colorimetric smartphone software applications that can determine a concentration of an analyte in a biological sample by conducting color analysis [12]. They work in a way similar to traditional colorimeters based on Beer–Lambert's law relating absorbance of a homogeneous medium to the concentration of the absorbing species [13]. Collimated monochromatic beam is irradiated to the medium and propagates the distance through the absorptive medium. Another approach is to use color analysis to determine a concentration of a colored substance. For example, there was a smartphone-based colorimetric reader for quantitative analysis of direct enzyme-linked immunosorbent assay (ELISA) for horse radish peroxidase, rapid sandwich ELISA for human C-reactive protein, and commercially available BCA (bicinchoninic acid) protein estimation assay [14]. In addition to that, color analysis can be used to measure the amount of free hemoglobin in plasma [15] and analyze color of dipsticks used for urinalysis [16].

3. Hardware requirements for the mobile health systems focused on rapid kits

Smartphone-based colorimeters are most commonly used for rapid kits toward mobile health systems as they can take very reproducible images and instantly conduct color analysis. They can broadly be classified into two categories: ones that are based on stand-alone applications, and others based on software combined with different hardware components. Mobile phone colorimeters of first kind are low-cost alternative for expensive commercial colorimetric readers [12]. The main limitation of them is that the user needs to recalibrate them with even a slightest change in ambient light.

Smartphones with an app combined with a dedicated light-controlling hoods and integrated lighting setup works great in changing ambient lighting. The light-controlling hood can also be used for holding a sample at constant position for continuous tracking of color change at particular location on the sample. Building a hood for a mobile colorimeter requires accurate fit to dimensions of the smartphone and position of the camera. In addition to this, its operation requires additional battery powered light source, for example, LED array or built-in flash light, to keep the illumination constant. Usually such attachments to smartphones are quite sophisticated, and assembly is challenging.

Sometimes, sample preprocessing requires additional preparation steps, for example, incubation at a specific temperature. In those cases, the attachment to the phone might be more sophisticated than just a light-controlling hood and/or sample mount. An incubator can be integrated in this portable system at low cost using Arduino, as a powerful electronics prototyping platform. This way complete functionality of ELISA can be integrated with a smartphone for less than $40. Later we discuss a number of these technologies in more detail.

4. Smartphone-based precise urinalysis

This project demonstrates how very simple attachment to a smartphone can greatly improve performance of urinalysis in comparison with the visual assessment [16]. Here, authors proposed a simple and low-cost hybrid point-of-care smartphone-based colorimeter (Fig. 6.1). It analyzes change in color of 10 sensing pads responsible for detection and quantitative measurement of different urine components (glucose, protein content, etc.). This system is more precise than an exclusively software-based colorimeter, but also does not require expensive or complex attachments, demonstrates great performance, high reproducibility, accuracy, and stability under varying lighting conditions. Additionally, it can be used for a very precise tracking of such critical indicators of preeclampsia and gestational diabetes, as concentration of protein and glucose in urine samples. Several important design considerations are made to allow the system to achieve high quality performance, which can potentially be used in different mobile health systems to increase accuracy and reproducibility of the results. They are described in several subsections.

4.1 Consistent imaging

One of the important steps toward achieving high assay reproducibility under various lighting conditions was to introduce a 3D-printed sample holder called Chroma-dock (Fig. 6.1C and D) that consists of a black box attached to a smartphone and a removable cassette serving as a sample holder. This structure not only allows keeping the sample at constant distance from the camera, but also creates a controlled-light environment effectively blocking all outside light. This allows capturing highly reproducible images and eliminates the background noise. The test

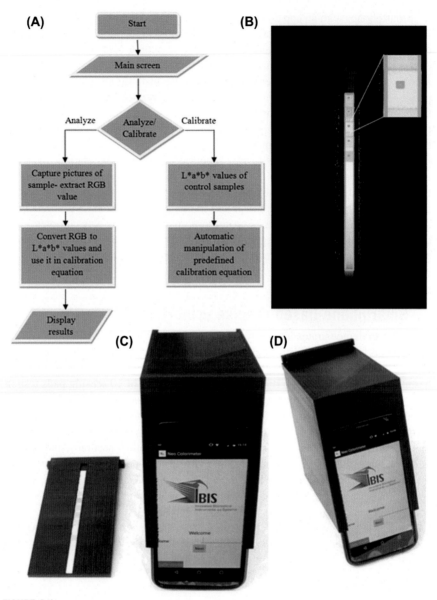

FIGURE 6.1

Smartphone-based urinalysis. (A) Diagram of the colorimetric analysis and calibration; (B) Camera preview with region of interest (ROI) over the reagent pads of the test strip; (C) Smartphone-based colorimeter with a cassette and a holder; (D) The cassette is inserted.

Reprinted from Ref. [16] with permission, © 2017 IEEE.

was conducted under bright light illumination, dim light, and in a very low light intensity, and the Chroma-dock allowed to take highly reproducible images independently of outside lighting conditions (Fig. 6.2). The distance between the camera and the test strip is ∼15 cm. The strip is placed into the groove at the center of cassette as shown in Fig. 6.1B. Then, the cassette is inserted into the black box that holds the strip in a fixed position during the color analysis. In addition to isolating sample from the surrounding light and fixing distance from the camera, there has to be a stable light source [17] and a number of camera control parameters that also have to be fixed to achieve better image reproducibility. As smartphone camera acts as a detector measuring color characteristics of the sample, the camera parameters such as exposure rate, white balance, sharpness, and ISO play a vital role in error-free colorimetric measurement. Exposure rate defines the amount of light per unit area that reaches the camera sensor. The colorimetric measurements were performed inside a black box where auto settings can cause overexposure and loss of color details [18]. To avoid this, the exposure compensation value was set to the minimum.

A white balance setting adjusts the color of the captured image based on the light source used while shooting the picture. Color reproducibility for images captured under different light conditions can be achieved by using the automatic white balance setting [19]. As the system was using a stable light source, the white balance was set to the constant mode. Specifically, the white balance was programmed in the smartphone application to use the parameters of the 'daylight' mode that allows to normalize the color values based on standard daylight illuminant source D65 (commonly used standard illuminant defined by the International Commission on Illumination).

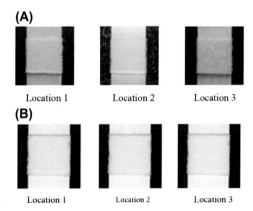

FIGURE 6.2

Comparison between the same test strips imaged in different locations with different ambient lighting conditions. (A) Images taken without Chroma-dock and (B) bottom three images were taken with Chroma-dock. Excellent repeatability is observed when Chroma-dock is used.

Reprinted from Ref. [16] with permission, © 2017 IEEE.

Autofocus is a critical feature for stand-alone smartphone colorimeters that use automatic focus at a specific ROI. The movement of the camera might cause a blur in the image that can add error to the measurements [20]. Constant distance between the object and the camera eliminates the need for autofocus. Therefore, the smartphone application was programmed to use 'fixed focus" mode. ISO number of a digital camera measures the sensitivity of the image sensor. The larger the ISO number, the worse is the signal-to-noise ratio (SNR) [21], so a lower ISO value (ISO 400) was programmed to yield a better SNR. In a case of a stand-alone application-based colorimeter that does not include any additional hardware, these settings have to be dynamically changed, according to varying environmental conditions—otherwise it introduces additional noise.

Finally, another important component of a stable smartphone colorimeter is a high-quality light source. Although an external LED can be used for smartphone colorimetric measurements, it would require an external power supply and wiring, which adds complexity to the system. Fortunately, contemporary smartphones have high quality built-in flashlight (white LED). This light source can be directly controlled using software installed on a smartphone.

4.2 Development of robust image processing software on a smartphone

Smartphone application functionality for colorimetric measurements involves capturing, storing, and analysis of an image of a sample. To determine a concentration of a substance, color information from an image has to be extracted and matched with the values from a calibration curve. Fig. 6.1A shows the algorithm used for the colorimetric analysis and calibration.

The image of the dipstick is taken using a built-in camera. After that, a ROI is chosen in the middle of a test pad responsible for the needed analyte. The ROI is a square 10×10 pixels. Color information in the ROI of the captured images as shown in Fig. 6.1B is extracted, and red, green, and blue (RGB) values of the corresponding pixels inside the ROI are obtained. RGB is a nonabsolute color space as the color values depend on external factors such as illumination and sensitivity of camera sensor [22−24]. CIE L*a*b* color model (a color space defined by the International Commission on Illumination) provides more accurate and uniform color representation [25]. L* value indicates lightness and it ranges from 0 to 100 (black to white). a* value indicates red/green color components (positive value represents red region and negative value represents green region). b* value indicates yellow/blue color components (positive value represents yellow region and negative value represents blue region). There is no direct standard formula to convert RGB to L*a*b* values. Smartphone app is programmed to convert obtained RGB values to CIE L*a*b* values indirectly, by calculating XYZ tristimulus values. Standard illuminant D65 was considered for the RGB to L*a*b* color space conversion.

4.3 Development of calibration algorithm

Color values obtained in the previous step were used to compute the concentration value of the substance. Equations fitting the calibration curves were built into the smartphone app. Some substances change color nonlinearly with linear change of concentrations, which adds complexity to the computation procedure. In this case, calibration equation of particular color component from L*, a*, and b* is used to determine the concentration of the substance in the sample and the values obtained from the calibration equations of other color components are used in further decision-making process.

5. Allergen detection

The next mobile health system is focused on rapid detection of allergens in food. It can be lifesaving in case of patients suffering from severe allergic reactions to food. Currently, 32 million people in United States have food allergies, including 5.6 million children under age 18 [26]. The symptoms of an allergic reaction to food can range from mild-to-severe throat tightening, difficulty breathing, anaphylaxis, and, in worse case, death. Coskun et al. proposed a smartphone-based solution for personalized allergen testing platform called iTube. The detection mechanism is based on sensitive colorimetric assay processed in test tubes comparing amount of food allergen in a food sample to one in a control sample. In addition to the smartphone, the system requires several low-weight and low-cost components: plastic plano-convex lens, two LEDs, two light diffusers, and circular apertures to spatially control the imaging field of view. The whole assembly beside the smartphone weighs only 40 g. The test and control tubes have reagent specific to an allergen to be detected, their image is processed on a smartphone within 1 s, and then the allergen concentration is reliably determined.

Although this approach can be used for a variety of allergens, the study [27,28] was focused on peanut detection using commercially available Veratox test kit, Neogen 8430. The sample preparation required grinding the food sample to fine particles, mixing it with hot water, and extraction of solvent. After that, three drops of solution without any solid content was used for the measurements. The color-producing step required rinsing with conjugate, substrate, and stop solution followed by wash buffer after each step. Overall, it took only 10 min to perform, and the color assessment on the smartphone provided quantitative information about presence of peanuts in the food sample.

In comparison with the first mobile health system (the one shown in Fig. 6.2), the attachment to the smartphone is a little more complicated while it provides more functionality. Although one of the functions is to isolate the test tubes from the outside lighting conditions and fix the distance between the samples and the camera, the optical system is designed differently. Instead of using flash light integrated on a smartphone, the system uses additional LED sources and some simple optical setup to properly control light propagation. It requires some minimal assembly but can be mass-produced at relatively low cost. The benefit of this approach in comparison with the first system is in an additional control over the light source. It can be monochromatic, and its color can be chosen based on a specific application, while the illumination with flash light is preferred when samples might have many different colors, or if they are not transparent, as in case of a dipstick urinalysis.

6. Smartphone-based detection of zika and dengue

Next smartphone-based system is focused on high accuracy testing of Zika and dengue [29] at low resource setting. The epidemics of Zika in 2015 and the terrible outcomes for pregnant women and their babies affected by this virus demonstrated the importance of prompt detection and differentiation of mosquito-transmitted diseases. Although Zika infections during pregnancy correlate with severe birth defects, including microcephaly and Guillain-Barré syndrome, there are no similar outcomes for dengue. Therefore, as a part of the prenatal care, it is really important to diagnose the correct virus. One of the challenges is that such flaviviruses as Zika and dengue can be difficult to differentiate because of their cross-reactivity in diagnostic tests. At the same time, more than a quarter of the world's population is at risk of dengue infection, and also hundreds of millions of people are infected annually. Therefore, wrongly diagnosed Zika for a pregnant woman affected by dengue can completely change outcomes of the pregnancy and lead to abortion of a healthy fetus.

The study published in Ref. [29] demonstrates that properly designed antibodies with good selectivity incorporated into convenient immunochromatography format for a color-based readout can be reliably used with a smartphone to accurately detect Zika and dengue. The sample used for the testing was 30−150 μL of serum. Sensitivity and specificity for detection of four dengue serotypes were between 0.76 and 1.0 (76%−100%), while for Zika it was 0.81 (81%) and 0.86 (86%), respectively. The biomedical tests are using monoclonal antibodies for detection of viral nonstructural 1 (NS1) protein antigens specific to dengue and Zika. The resulting assay can be used for identification of four dengue serotypes and Zika viral infections without cross-reaction, and the smartphone allowed to conduct these tests with much higher accuracy that can be achieved with just visual assessment.

7. Rapid blood plasma analysis for detection of hemolysis

The color characterization on a smartphone can be used even without any color changing reagents. One of the examples of such mobile health system is the hemolysis sensor designed for detection of preeclampsia and HELLP (hemolysis, elevated liver enzymes, and low platelet count) syndrome: pregnancy-related complications with high rates of morbidity and mortality [15]. HELLP syndrome, in particular, can be difficult to diagnose. Recent work suggests that elevated levels of free cell hemoglobin in blood plasma can, as early as the first trimester, potentially serve as a diagnostic biomarker for impending complications. Here, authors developed a point-of-care smartphone-based platform that can quickly characterize a patient's level of hemolysis by analyzing the color of blood plasma (Fig. 6.3) [15]. The custom hardware and software are designed to be easy to use. A sample of the whole blood (~10 μL or less) is first collected into a clear capillary tube or microtube, which is then inserted into a low-cost 3D-printed sample holder attached to the phone. A 5–10 min period of quiescence allows for gravitational sedimentation of the red blood cells, leaving a layer of yellowish plasma at the top of the tube.

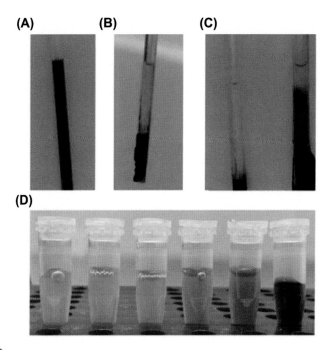

FIGURE 6.3

Plasma separation and colorimetric analysis of free hemoglobin concentrations. (A) Human whole blood is first collected into a thin capillary tube (1 mm diameter). (B) Red blood cells are then allowed to settle gravitationally to the bottom of the tube. The top layer (clear color) is isolated plasma. (C) Different levels of hemolysis (i.e., concentrations of hemoglobin) in the plasma are detectable as color differences. The sample on the right is hemolyzed; the sample on the left is not. (D) Solution series showing a range of concentrations of free hemoglobin in plasma in microtubes, from 4 mg/dL (leftmost sample) to 300 mg/dL (rightmost sample).

Reprinted from Ref. [15] with permission, © 2017 Elsevier.

The phone camera then photographs the capillary tube and analyzes the color components of the cell-free plasma layer. The software converts these color values to a concentration of free hemoglobin, based on a built-in calibration curve, and reports the patient's hemolysis level: nonhemolyzed, slightly hemolyzed, mildly hemolyzed, frankly hemolyzed, or grossly hemolyzed. The accuracy of the method is ~1 mg/dL. This phone-based point-of-care system provides the potentially life-saving advantage of a turnaround time of about 10 min (vs. 4+ hours for conventional laboratory analytical methods) and a cost of approximately $1 (assuming the user have a smartphone and the software is available).

8. Diagnosis of infectious diseases at the point of care using ELISA-like assays

Patient studies with smartphone-based systems integrated with rapid kits demonstrate strong preference over traditional laboratory-based testing. For example, another study published in Ref. [30] evaluated on 96 patients in Rwanda showed 97% preference because of quick results obtained with a single finger prick. This study used a small and light attachment to the smartphone, called a "dongle" that ran assays on disposable plastic cassettes with preloaded reagents. The disease-specific zones were used for readout similar to an ELISA microplate assay. The assay itself was also similar to ELISA, but gold nanoparticles and silver ions were used for the amplification step instead of enzymes and substrate. The assay was detecting biomarkers of HIV and syphilis in pregnant women to minimize disease transmission to their children. The field studies demonstrated that even under much more challenging conditions than conducting testing in controlled laboratory environment, a 15-min assay using the dongle provided accurate diagnostic results on triplexed markers for HIV and syphilis.

The high-accuracy ELISA-like system required fabrication of the attachment to the phone and a custom-made reagent cassette. The dongle was made using custom-printed circuit boards, LEDs and photodiodes, and 3D-printed casing. Vacuum chamber was created with a one-way umbrella valve, a rubber bulb from a 140-mL syringe, and a conical spring. The cassettes were prepared at Columbia University before transporting to Rwanda. The disease-specific proteins were added to the cassette surface by direct physisorption with a stabilizing agent or covalently attached to the plastic surface using EDC-sulfo-NHS reaction (water-soluble 1-ethyl-3-[3-dimethylaminopropyl] carbodiimide (EDC), N-hydroxysulfosuccinimide(Sulfo-NHS)). The cassettes were prepared using robot-assisted manufacturing for better reproducibility and higher throughput. Overall, the complete system is much more complicated in design than a simple box attached to a smartphone, but it still satisfies all requirements for real world applications, as the complete attachment can be produced for just $34.

9. Complete replication of ELISA functionality on a smartphone

Although development of ELISA-like assays with custom-made cassettes described in the previous section has a lot of practical value, development of a system that is capable to replicate complete functionality of traditional ELISA and use all standard reagents available in thousands of commercial kits, would be extremely useful. Overall, ELISA is one of the most important technologies for biochemical analysis critical for diagnosis and monitoring of many diseases. Traditional systems for ELISA incubation and reading are expensive and bulky, thus cannot be used at point of care or in the field. Here, we proposed and demonstrated a new miniature smartphone-based system for ELISA [31]. This system can be used to complete all steps of the assay, including incubation and reading. It weighs just 1 pound, can be fabricated at low cost, portable, and can transfer test results via smartphone. It was also shown how mobile ELISA can be calibrated for accurate measurements of progesterone and demonstrated successful measurements with the calibrated system. Fig. 6.4 shows the attachment to the smartphone that can be fabricated for under $35. The incubation and reading are conducted inside of a dark box. The standard ELISA microwells are initially incubated at 37°C, then they are washed and all other standard ELISA steps are conducted. At the end the microwells are transferred to an imaging tray, and their color is analyzed using a smartphone. For the production, the box and the trays are 3D-printed. All electronics, including sensors, heaters, and LCD screen used as a backlight, are controlled using Arduino. All components are available at low cost and can be easily assembled into a functional system.

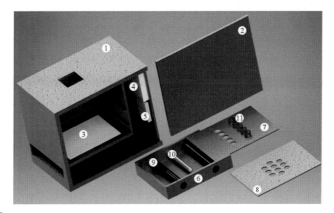

FIGURE 6.4

3D model of the complete system (artistic rendering). Labeled components: (1) prototype housing, (2) sliding door, (3) LCD backlight screen and Arduino Mega 2560, (4) circuit board, (5) DC solid state relays, (6) water bath, (7) heating tray, (8) image capturing tray, (9) copper tubes with heaters, (10) temperature probe, and (11) ELISA microwells.

Reprinted from Ref. [31] with permission, © 2018 Elsevier.

10. Sperm testing for PC and smartphones

The final example of a rapid kit smartphone system is also based on image processing but does not use color analysis. The system is focused on sperm testing for fertility assessment [11]. Currently, more than 100 million couples worldwide are unable to conceive a child, and one of the major reasons for that is that male infertility affects up to 12% of the world's male population. The sperm analysis is one of the important diagnostic tools, unfortunately available at relatively high cost in traditional laboratories. As a result, it cannot be used in lower income countries, or at home.

The smartphone-based system for sperm analysis can be privately used at home and is low cost enough for the resource limited settings. For the evaluation, the authors used 350 clinical semen specimens at a fertility clinic and demonstrated that their assay can analyze an unwashed, unprocessed liquefied semen sample with <5 s mean processing time and provide the user a semen quality evaluation based on the World Health Organization (WHO) guidelines with ~98% accuracy. The system was based on an optical attachment for microscopic imaging on a phone, and a disposable microfluidic device for semen sample handling. The disposable microfluidic device for semen sample handling was designed on the basis of a power-free mechanical pumping mechanism using a polydimethylsiloxane bulb for easy-to-use, on-chip semen sample loading through creating a negative pressure chamber at the end of the device. The total material cost of the additional components was $4.45, including $3.59 for the optical attachment and $0.86 for the microfluidic device. The videos were taken at a rate of 30 frames per second (fps) with a maximum effective FOV (Field of view) of 886 × 886 pixels, and the results were reported at an average time of 4.48 s. Although the described system is published in Ref. [11] and presented as a prototype, there is an FDA approved, commercially available smartphone-based system for sperm testing. The YO Sperm Tests provides automatic results and a live video of moving spermatozoids [32].

11. Conclusion

Smartphones are a great platform for running a variety of rapid tests. The power of this approach is in low cost and wide availability of smartphones and low cost of integration of detection elements into a sensing system. Therefore, this approach is preferred when biomedical tests have to be conducted in resource limited settings, at low cost, but without loss of accuracy comparing to gold standard technologies. However, to reach its full potential, the system has to be carefully designed for maximum accuracy comparable with the available gold standards, and variety of additional functional elements. Many of those additional components can be mass produced or prototyped using low cost technologies, such as 3D printers, and inexpensive optical, mechanical, and microfluidic components. Although mobile health is still relatively young research area, success of several field patient studies, and even FDA approval of some of the systems, clearly indicate, that it is going to develop significantly further, and that we should expect many more useful products.

References

[1] Information and communications technology facts and figures — the World in 2015. International Telecommunication Union; 2015.

[2] Tsai CC, et al. Usability and feasibility of PmEB: a mobile phone application for monitoring real time caloric balance. Mob Netw Appl 2007;12(2—3):173—84.

[3] Tseng D, et al. Lensfree microscopy on a cellphone. Lab Chip 2010;10(14):1787—92.

[4] Logan AG, et al. Mobile phone—based remote patient monitoring system for management of hypertension in diabetic patients. Am J Hypertens 2007;20(9):942—8.

[5] Giavazzi F, et al. A fast and simple label-free immunoassay based on a smartphone. Biosens Bioelectron 2014;58:395—402.

[6] Arora S, et al. Detecting and monitoring the symptoms of Parkinson's disease using smartphones: a pilot study. Park Relat Disord 2015;21(6):650—3.

[7] Bourouis A, et al. An intelligent mobile based decision support system for retinal disease diagnosis. Decis Support Syst 2014;59:341—50.

[8] Kwon H, et al. A smartphone metabolomics platform and its application to the assessment of cisplatin-induced kidney toxicity. Anal Chim Acta 2014;845:15—22.

[9] Zhu H, et al. Optofluidic fluorescent imaging cytometry on a cell phone. Anal Chem 2011;83(17):6641—7.

[10] Preechaburana P, et al. Biosensing with cell phones. Trends Biotechnol 2014;32(7):351—5.

[11] Kanakasabapathy MK, Sadasivam M, Singh A, Preston C, Thirumalaraju P, Venkataraman M, Bormann CL, Shehata Draz M, Petrozza JC, Shafiee H. An automated smartphone-based diagnostic assay for point-of-care semen analysis. Sci Transl Med 2017;9(382):eaai7863.

[12] Yetisen AK, et al. A smartphone algorithm with inter-phone repeatability for the analysis of colorimetric tests. Sens Actuators B Chem 2014;196:156—60.

[13] McNaught AD, McNaught AD. Compendium of chemical terminology, vol. 1669. Blackwell Science Oxford; 1997.

[14] Vashist SK, et al. A smartphone-based colorimetric reader for bioanalytical applications using the screen-based bottom illumination provided by gadgets. Biosens Bioelectron 2015;67:248—55.

[15] Archibong E, Konnaiyan KR, Kaplan H, Pyayt A. A mobile phone-based approach to detection of hemolysis. Biosens Bioelectron 2017;88:204—9.

[16] Konnaiyan KR, Cheemalapati S, Gubanov M, Pyayt A. mHealth dipstick analyzer for monitoring of pregnancy complications. IEEE Sens J 2017;17(22):7311—6.

[17] Caglar A, et al. Could digital imaging be an alternative for digital colorimeters? Clin Oral Investig 2010;14(6):713—8.

[18] Hirsch R. Exploring color photography fifth edition: from film to pixels. Taylor & Francis; 2013.

[19] Hsu E, et al. Light mixture estimation for spatially varying white balance. In: ACM transactions on graphics (TOG). ACM; 2008.

[20] Brown G. How autofocus cameras work. 2008. Internet Article, http://travel. howstuffworks.com/autofocus.htm.

[21] Foi A, et al. Practical poissonian-Gaussian noise modeling and fitting for single-image raw-data. Image process IEEE Trans 2008;17(10):1737—54.

[22] Mendoza F, Aguilera J. Application of image analysis for classification of ripening bananas. J Food Sci 2004;69(9):E471—7.

[23] Segnini S, Dejmek P, Öste R. A low cost video technique for colour measurement of potato chips. LWT-Food Sci Technol 1999;32(4):216—22.

[24] Paschos G. Perceptually uniform color spaces for color texture analysis: an empirical evaluation. Image Process IEEE Trans 2001;10(6):932—7.

[25] Korifi R, et al. CIEL*a*b* color space predictive models for colorimetry devices— Analysisof perfume quality. Talanta 2013;104:58—66.

[26] https://www.foodallergy.org/life-with-food-allergies/food-allergy-101/facts-and-statistics.

[27] Coskun AF, Wong J, Khodadadi D, Nagi R, Tey A, Ozcan A. A personalized food allergen testing platform on a cellphone. Lab Chip 2013;13(4):636—40.

[28] Coskun AF. Lensfree fluorescent computational microscopy on a chip (Ph.D. diss.). UCLA; 2013.

[29] Bosch I, de Puig H, Hiley M, Carré-Camps M, Perdomo-Celis F, Narváez CF, Salgado DM, et al. Rapid antigen tests for dengue virus serotypes and Zika virus in patient serum. Sci Transl Med 2017;9(409):eaan1589.

[30] Laksanasopin T, Guo TW, Nayak S, Sridhara AA, Xie S, Olowookere OO, Cadinu P, et al. A smartphone dongle for diagnosis of infectious diseases at the point of care. Sci Transl Med 2015;7(273):273re1.

[31] Zhdanov A, Keefe J, Franco-Waite L, Konnaiyan KR, Pyayt A. Mobile phone based ELISA (MELISA). Biosens Bioelectron 2018;103:138—42.

[32] https://www.yospermtest.com/. Medical Electronic Systems.

Smartphone-based medical diagnostics with microfluidic devices

Dong Woo Kim[1], Kwan Young Jeong[1], Hyun C. Yoon, PhD[2]

[1]*Department of Applied Chemistry & Biological Engineering and Department of Molecular Science & Technology, Ajou University, Suwon, Republic of Korea;* [2]*Professor, Department of Applied Chemistry and Biological Engineering, Department of Molecular Science and Technology, Ajou University, Suwon, Republic of Korea*

ABBREVIATIONS

ELISA	enzyme-linked immunosorbent assay
POCT	point-of-care testing
HRP	horseradish peroxidase
TMB	3,3′,5,5′-tetramethylbenzidine
LCD	liquid crystal display
RGB	red, green, blue
PMMA	polymethylmethacrylate
PDMS	polydimethylsiloxane
PAD	paper-based analytical device
RBC	red blood cell
LFIA	lateral flow immunoassay
GNP	gold nanoparticle
MAOS	N-Ethyl-N-[2-hydroxy-3-(sodiosulfo)propyl]-3,5-dimethylaniline
4-AAP	4-aminoantipyrine
NFC	near field communication
LOD	limit of detection
GOx	glucose oxidase
CMOS	complementary metal—oxide—semiconductor
PCR	polymerase chain reaction
LAMP	loop-mediated isothermal amplification
RPA	recombinase polymerase amplification
FACS	fluorescence-activated cell sorting
HIV	human immunodeficiency virus
LED	light emitting diode
PET	polyethylene terephthalate
TMBM	3,3′,5,5′-tetramethylbenzidine membrane
AgNPRs	triangular silver nanoprisms
CEA	carcinoembryonic antigen
ATP	adenosine triphosphate
LSPR	localized surface plasmon resonance

Smartphone Based Medical Diagnostics. https://doi.org/10.1016/B978-0-12-817044-1.00007-7

CTX-II	carboxy-terminal telepeptides of type II collagen
OLED	organic light-emitting diode
USB	universal serial bus
TCNQ	tetracyanoquinodimethane
TTF	tetrathiafulvalene
TPA	tripropylamine
ECL	electrochemiluminescence
RJP	retroreflective Janus particle
cTnI	cardiac troponin I
PTC	positive temperature coefficient
QR	quick response
CPU	central processing unit
GPS	global positioning system
Wi-Fi	wireless fidelity
LTE	long-term evolution
RFID	radio frequency identification

1. Introduction

For predicting or diagnosing diseases, patients have been admitted to laboratories and hospitals, and their biological specimens (blood, urine, swabs, etc.) were assays using large equipment. This equipment provides accurate, sensitive, and reliable assay results necessary for making clinical decisions. However, they suffer from high cost and long assay time. Recently, point-of-care diagnostics devices have become increasingly popular, which can be used in individual doctors' offices and at home. Moreover, for infectious and fatal diseases such as MERS, ZIKA fever, and Ebola, early detection and isolation of infected individuals are crucial. Therefore, the diagnostic paradigm is changing from a central clinical laboratory (where large and high-end medical diagnostics devices are available) to a smaller laboratory or an individual doctor's office (using small-scale devices). In response to this paradigm shift, many researchers in both academia and industry have developed numerous small-scale diagnostics devices with performance similar to large equipment. In particular, the development of microfluidic systems integrated to the lab-on-a-chip platform has become a key technology enabling miniaturization of such diagnostic equipment [1−3].

This microfluidic system allows for reducing the device size, reagent consumption, and assay time. Moreover, multiple and/or complicated diagnostics can be performed on a single microfluidic chip, eliminating the need for complex procedures and saving time/cost. These microfluidic systems are useful in developing and underdeveloped countries where infrastructure is insufficient and access to clinical laboratory is limited. However, despite numerous efforts, there remain several challenges. A skilled operator is often required because conventional signal transducers are made of complicated external instruments. Furthermore, the current size of microfluidic diagnostic devices remains on a benchtop scale and unsuitable as portable instruments due to bulky signal transducers. Therefore, a portable transducer is needed for microfluidic diagnostic systems. To overcome this limitation, many researchers have developed and reported the use of the smartphone as a signal transducer. As the smartphone has a small size, the smartphone-based diagnostic platform is suitable as a portable medical diagnostic device [4—6]. In addition, smartphone-based diagnostics can be easily used by end user and eventually patients at the point of care or even at home because most people are familiar with smartphones. Above all, current smartphones are loaded with high-end technical components, such as a high-resolution camera module, flashlight, illumination sensor, large capacity battery, a wireless networking module, and an arithmetic unit, and these components are highly competitive to the components used in commercial medical diagnostic devices [7—9].

Most components in smartphones as depicted in Fig. 7.1 offer high accuracy, sensitivity, and portability. Therefore, there is a great potential for using smartphone-embedded components as signal transducers for a medical diagnostic device. In this chapter, we will discuss how each smartphone-embedded component can be utilized for biosensing and medical diagnostics from microfluidic systems.

2. Medical diagnostics using microfluidics and the camera module of the smartphone

Among the aforementioned various components of the smartphone that can be applied to the medical diagnostic device, those using the complementary metal—

FIGURE 7.1

Schematic illustration of the smartphone-embedded components that could be applied to the biosensor.

oxide–semiconductor (CMOS) camera of the smartphone have been actively researched and developed. The analytical device using the smartphone-embedded camera has the advantages of the optical analysis method, such as providing high sensitivity and intuitive results. In addition, in the case of requiring a light source, the smartphone-embedded flash can be used autonomously as a light source without an additional light-emitting device. Moreover, as the performance of the smartphone-embedded camera has been improved with the development of technology, obtaining high-resolution images similar to those obtained using large equipment can be achieved. Therefore, it is possible to miniaturize the equipment by applying the smartphone camera module to the medical device for diagnosis. In this section, we will describe the optical diagnostic devices that integrate with the microfluidic system based on the smartphone-embedded camera module using colorimetric, fluorescence, retroreflective, and microscopic bioimaging.

2.1 Detection of the colorimetric signal using the smartphone-embedded camera

Among the smartphone-based optical diagnostic methods, colorimetric methods are widely used. In the colorimetric assay, there are two methods of measuring the colorimetric reaction: measuring transmitted light or measuring reflected light. In measuring transmitted light, the light transmitted through the colored solution is detected by smartphone camera, and the amount of detected light decreases with the degree of color generation. When measuring reflected light, the light illuminates the substrate where the color generated, and the reflected light of the substrate is observed by the smartphone camera. These two methods are widely used to detect a colorimetric reaction by measuring absorbance and reflectance, respectively.

A typical example of measuring the transmitted light in the colorimetric method is in the enzyme-linked immunosorbent assay (ELISA), which is used for measuring the concentration of biochemical and biomarkers using enzyme-mediated colored product generation and is commercially available as a benchtop-size apparatus. However, as it is difficult to use benchtop-size ELISA equipment as a point-of-care testing (POCT) device, smartphone-based portable ELISA instruments were studied [10,11]. In the developed system, the light source illuminates the oxidized 3,3',5,5'-tetramethylbenzidine (TMB) by the horseradish peroxidase (HRP)-conjugated antibody in the sample reservoir. The oxidized TMB absorbs the light corresponding to its maximum absorption wavelength (652 nm), and the transmitted light reaches the smartphone camera through an additional lens and aperture in the optical cradle. The incident light on the smartphone camera is analyzed and calculated as an alternation of absorbance, or the ratio between the incident light and transmitted light. In other words, the absorbance change can be obtained by measuring the transmitted light passing through the chromogenic sample, and the amount of antigen can be calculated by the difference in absorbance.

In the proposed manuscripts, an accurate quantitative analysis is possible, but additional optical components such as monochromatic light and another light source are still required. These additional optical components which are attached to smartphone cause increasing the weight of gadget, poor user friendliness, low cost-effectiveness, and high complexity.

To overcome these limitations, Chun et al. developed a simple optical biosensing platform using the principle of light filtering instead of the spectroscopic method for analysis of transmitted light [12]. The principle of specific wavelength filtering is similar to a secret message card. The secret message card set consists of a word-printed card and the semitransparent cellophane film. Although an observer cannot see the printed word, which has the same color as the cellophane film due to the light interference effect, the words with a different color than that of the cellophane film can be observed through the cellophane. Using this principle, as depicted in Fig. 7.2A, the cellophane filter is replaced with the blue-colored microfluidic channel produced by the oxidized TMB, and the printed word is replaced with the signal guide, which has various colored polygonal figures in the LCD monitor, respectively. In the microfluidic channel, when more target analyte is present, the colorimetric reaction that makes the blue-colored product is activated. As the microfluidic channel, which has more colored products, obscures more of the polygonal figures, it is possible to measure the target by only recognizing the number of each figure. In resource-limited conditions, the proposed smartphone-based colorimetric biosensing platform will be able to simply measure the target molecule such as glucose, a biomarker that is user friendly and intuitive.

Rather than measuring the transmitted light to detect the absorbance, a simpler method is to measure the change of the reflected intensity of the analyte—reagent

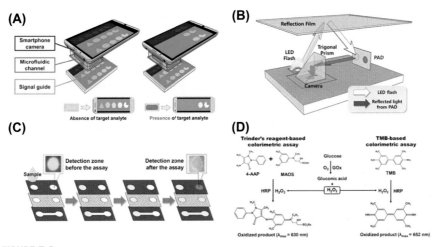

FIGURE 7.2

(A) Schematic illustration of colorimetric assay proposed by Chun et al. (B, C) Schematic illustration of paper-based colorimetric assay proposed by Chun et al. (D) Mechanism of a colorimetric assay for glucose detection.

complex. Various studies describing the measurement of reflected light from microfluidics using a smartphone-embedded camera have been published [13,14]. For reflected color measurement using the smartphone-embedded camera, a transparent microfluidic chip is preferred. If materials have a color that is incompatible with the sample, it is difficult to obtain accurate results due to the color interference phenomenon caused by the materials. Therefore, extensive research has been performed to measure the reflected color intensity on a transparent polymer microfluidic chip such as PMMA and PDMS. To fabricate the microfluidic device with transparent polymers, lithography techniques were generally used [15]. In addition to these lithographic techniques, due to advances in 3D printing technology, complex microfluidic devices have been simply fabricated using 3D printers [16–18]. The advantage of the 3D-printed microfluidic chip is that it is possible to implement various features such as a simple pump, solution mixer, and various types of valves [19]. However, the 3D-printed microfluidic chip has limitations such as poor resolution, extra work to remove supporting material, and the roughness of channel that may cause uneven light scattering. Despite these drawbacks, as the 3D-printed microfluidic chip has greater cost-effectiveness than lithography techniques, it has a potential for expanding into a wide field.

In perspective of commercialization, a paper-based analytical device (PAD) has more suitable than polymer-based microfluidic chip because the PAD presents several advantages; it can be inexpensive, easy to store, and easy to move. In addition, it can be operated in the absence of an external power pump and it can filter out various particles, RBCs in blood, or unwanted substances [20,21]. The fabrication methods of PAD are also simple and suitable for mass-production [22].

As a method that has a similar advantage to PAD, smartphone-based lateral flow immunoassay (LFIA) has been studied and commercialized for a long time. The LFIA reader consists of two main components, the light detector and the light source, which could be replaced with a camera module and a flashlight in the smartphone, respectively. Lee et al. published an article describing an LFIA detection method using a smartphone-based reading system, which consisted of a macro lens, white LED, and smartphone-embedded camera [23]. The principle of the LFIA involves measuring the total reflectance of the strip [24]. When the light is emitted from the white LED, and the light illuminates the test line, the colloidal gold nanoparticles (GNPs) that form the immunocomplex for the antigen in the test line absorb the light, and the intensity of the light reaching the light detector is decreased. Therefore, the more target antigen present in the sample, the greater the amount of light absorbed by the GNPs in the test line. Because the amount of light absorbed in the test line increases, the amount of light entering the light detector decreases. However, there is the disadvantage that such a single-strip LFIA cannot be used in a multiplexed assay [25]. After the hydrophobic-patterned paper microfluidic diagnostic platform was reported, a paper-based multiplexed diagnostic was made possible. As an example, Dong et al. conducted an immunoassay with a paper-based microfluidic device that has three channels using GNPs [26].

In addition to detecting the plasmonic phenomena of GNPs in the PAD, the colorimetric reaction can also be measured on the PAD with the smartphone gadget depicted in Fig. 7.2B [27,28]. When a sample solution is applied to a PAD, as shown in Fig. 7.2C, the sample solution moves to the hydrophilic region as a result of capillary force. After the sample solution reaches the region, where the coloring reagent is dropped, the color is changed because of the biochemical reaction. If glucose is the target to be analyzed, glucose is oxidized by GOx, producing gluconic acid and H_2O_2 (Fig. 7.2D). The HRP uses this H_2O_2 in the color reaction with MAOS and 4-AAP to produce a blue color. To measure the color change of the PAD, the image is obtained with the smartphone-embedded camera, which is used as a transducer for converting a color reaction into an image, and the RGB value is measured with computer software (NIH ImageJ). However, the limitation of measuring reflected light in the color reaction is that the images are often changed according to the external environment. Therefore, an external gadget that serves to block external light needs to be introduced to reduce this error. Moreover, to maintain a constant light condition, the flashlight of the smartphone was used [27−29]. However, as these additional gadgets have reduced portability, many studies have been performed to make measurements in ambient light conditions. Shen et al. showed that the color change induced by the pH change in the urine test strip could be measured with a mobile phone [30]. In addition, Koh et al. introduced reference color markers (true white and black) to enable white balancing according to light conditions (daylight, shadow, and various light conditions) [31]. Therefore, it was shown that substances existing in sweat such as glucose and lactate could be measured by the color reaction even under various light conditions.

In the diagnosis based on chromogenic reaction using the smartphone, there are advantages such that it requires the simple fabrication of a microfluidic device and measurement process. Therefore, colorimetric diagnosis using a smartphone will be commercialized as a next-generation diagnostic platform.

2.2 Smartphone-based fluorescence diagnostic system

Along with the colorimetric detection method, the fluorescence-based detection method is also commonly used in medical diagnosis. The immunodiagnostic methods using a fluorescent probe (fluorophore-impregnated beads, quantum dots, etc.) have the advantage of sensitivity, implying that they can be used to measure low-level targets. Therefore, a lower limit of detection (LOD) value is accomplished in fluorescence-based detection compared with colorimetric-based diagnostics [32].

As an example of fluorescence-based diagnosis, Coskun et al. developed the immunosensing platform in a smartphone for detecting albumin in urine, which is one of the biomarkers for kidney diseases, using fluorescent probe [33]. Generally, the smartphone-based fluorescence detection system consisted of a smartphone-embedded CMOS camera as the light detector, an external laser diode as the light source, an emission filter, and an additional lens, as illustrated in Fig. 7.3. This optics system is built into the gadget (albumin tester) that includes a microfluidic channel to

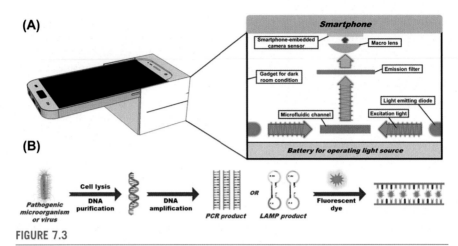

FIGURE 7.3

(A) Schematic illustration of smartphone-embedded fluorescence detection device. (B) Mechanism of pathogen detection using gene amplification methods and fluorescent dye.

detect fluorescent signals in darkroom conditions. When the laser light is irradiated onto the microfluidic channel, the fluorescent probe in the microfluidic channel is excited. The excited state fluorescent probe emits fluorescence while changing the energy level to the ground state. The emitted light is filtered by the emission filter, passed through the additional macro lens, and captured as an image by the smartphone camera. The fluorescence intensities in obtained images were analyzed using analysis software. Fluorescence intensity has a corelation with urinary albumin. Based on this principle, antigens and biochemicals can be quantitatively analyzed using the smartphone-based fluorescence detection platform in the polymer-based microfluidic system and PAD system [34—36]. In addition, using this smartphone-based fluorescence detection system, Rogers' group collected released sweat using the microfluidic system and measured the ions in the sweat, which is a clinical parameter, using a fluorescent probe. The ion measurements in sweat also can be performed in paper-based microfluidic systems as well as polymer-based microfluidic systems [37]. In the proposed fluorescence detection platform, the laser diode was utilized for providing excitation light of the correct wavelength. However, the heavy battery for operating the laser diode is not suitable for POCT in resource-limited conditions. Therefore, to eliminate the additional battery, which is a heavy component in the fluorescence detection gadget, a smartphone-embedded flashlight combined with an excitation filter was used as an excitation light [38]. In this case, a monochromatic laser for excitation was not required, but an excitation filter was required to illuminate only the monochromatic light because the flashlight of the smartphone emits multicolored white light. Although these have drawbacks, the fluorescence-based diagnosis widely is used and studied in POCT device due to high sensitivity.

The fluorescence-based diagnostic methods are used not only in immunoassays and biochemical detection but also to diagnose infections including those caused by pathogens such as *Salmonella* and *S. aureus* in the body. Although the pathogens can be detected using the same fluorescence-based immunoassays, the better and more accurate method is the use of gene amplification technology. A typical example of gene amplification technology is polymerase chain reaction (PCR), which is a technique for amplifying a specific gene fragment to multiple copies using polymerase and primer. As depicted in Fig. 7.3B, when the target-specific gene conserved in the pathogen was amplified using PCR, the amplified DNA fragment was present in the double-stranded DNA form, which could be then intercalated by fluorescent dye such as SYBR green or EVA green. Therefore, if the pathogen is present, the DNA fragment of the pathogen is amplified and detected as a fluorescence signal. Based on this principle, Snodgrass et al. developed a handheld PCR capable microfluidic device to measure KS herpesvirus [39]. Conventional benchtop-size PCR equipment can use a cooler, heating system, and thermocontroller for temperature control. However, handheld-type PCR devices that can be implanted in a smartphone cannot have these thermo-control components. Therefore, to facilitate temperature control, the microfluidic control device was divided into three sections in which temperature was maintained by the light irradiating such as sunlight or LED light. The target-specific gene from the pathogen was amplified by repeatedly passing through each section. After the amplification, when fluorescence dye (SYBR green) is intercalated to the dsDNA-form amplicon, a fluorescence signal according to the amount of amplified dsDNA can be obtained. The smartphone-embedded camera can be used to detect the fluorescence signal with an additional optical setup as depicted in Fig. 7.3A, and the obtained fluorescence signal can be analyzed using the analytical program. Using PCR and the fluorescence detecting method with the smartphone, sensitive and selective analysis of KSHV was accomplished. However, PCR-based diagnosis technology needs the thermo-controlling part for controlling the temperature for each step, that leads the increasing size of the diagnostic device and, consequently, weakening the portability.

To solve this problem, isothermal amplification techniques such as recombinase polymerase amplification (RPA) and loop-mediated isothermal amplification (LAMP), which can selectively amplify genes without temperature changes, have been spotlighted in molecular diagnostics [40]. Chen et al. reported the multiplexed detection of pathogens using the LAMP technique and smartphone [41]. In this study, a smartphone connectable handheld cradle that can be inserted into a multiplexed microfluidic chip was developed. The microfluidic chip has a sim card size of 25 by 15 mm and is fabricated using the photolithographic fabrication method. The microfluidic chip has one inlet and ten parallel assay channels. Each assay channel has a deposited primer for the specific pathogen-specific DNA sequence. When the sample enters each assay channel through the inlet and LAMP reaction occurs in the assay channel, the DNA of the target pathogen is amplified specifically in each channel. For measuring the concentration of amplicon, fluorescence dye was used. For excitation of fluorescent dye, a 485 nm blue LED combined with 490 nm short pass filter as an excitation filter was employed. The excited fluorescent dye emits red-shifted light while changing the energy level to the ground state. The emitted light was received by the smartphone camera through a 525-nm-long pass filter as an emission filter and additional macro lens. The fluorescence signal observed by the smartphone camera module was increased along with the concentration of amplified genes. By measuring the difference in intensity of fluorescence depending on the amount of target pathogen, it is possible to confirm the presence of pathogens as an end-point assay. In this apparatus, using a PTC heater connected to the 9 V battery, a temperature of 60−65°C for the LAMP reaction is maintained to sustain the activity of the *Bst* polymerase. Different from PCR devices, the equipment required for the isothermal temperature condition control is relatively simple. This electric heating system using a battery could be replaced by other heating systems such as a chemical heating system using a chemical reaction that is dissipated by Mg−Fe alloy, a phase-change material [42]. With these efforts, the isothermal gene amplification technology which combined with the smartphone-based diagnosis will become an indispensable technology for pathogen detection.

Despite attempts to simplify the equipment, the smartphone-based fluorescence measurement equipment still requires an emission filter, an external light source with monochromatic light, a built-in flashlight and excitation filter, and a dark room condition for blocking external light. Therefore, despite the sensitivity of the fluorescence detection system, miniaturization and cost problems need to be addressed for its use as a POCT device integrated into a smartphone.

2.3 Smartphone-based simple optical diagnostic system: retroreflective signaling probe

The fluorescence method has high sensitivity but still requires a sophisticated and complex optical system. Furthermore, a short fluorescence duration makes it difficult to detect the signal. These drawbacks have restricted the development of a smartphone-based POCT device. Therefore, the use of nonspectroscopic methods, which are less complex optical systems, can help make the smartphone-based diagnostic device a true POCT handheld device. In this section, we will cover the nonspectroscopic biosensor based on the retroreflection phenomenon.

The retroreflection phenomenon is a type of reflection in which the reflected light is redirected back to the incident direction. The retroreflection phenomenon occurs regardless of the light source at all wavelengths including visible light, infrared light, ultraviolet (UV) light, and nonmonochromatic light (white LED). The materials that induce the retroreflection are called retroreflectors. As depicted in Fig. 7.4A, there are two general types of retroreflective geometry. One is spherical as known as cat's eye type, and the other is a corner cube type [43]. The intensity of the retroreflected light depends on the retroreflector geometry. The corner cube types have a higher intensity than the spherical types, but this depends on the azimuth angle of the incident light. However, the spherical types cause constant retroreflection regardless of the azimuth angle of the incident light. Moreover, the spherical types are more accessible to manufacture than the corner cube types. Therefore, for using spherical-type retroreflectors as the biosensing probe, the retroreflective Janus particle (RJP) that is half metal-

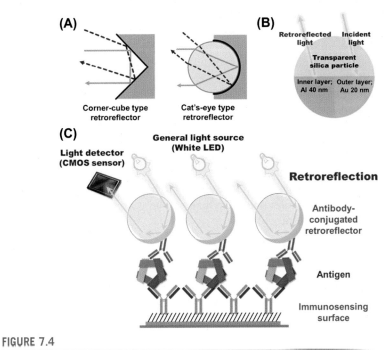

FIGURE 7.4

(A) Types of retroreflectors; corner-cube-type and cat's eye-type retroreflector. (B) Schematic illustration of metal half-coated silica particle as retroreflective biosensing probe; retroreflective Janus particle (RJP) developed by Han et al. (C) Schematic illustration of retroreflective biosensing platform proposed by Han et al.

coated 1200-nm-diameter silica particle with aluminum and gold, was developed in Yoon's group, as shown in Fig. 7.4B [44]. To use the RJP as a biosensing probe with the microfluidic system, the capture antibody and detection antibody were modified in sensing the surface and gold region of the RJP, respectively. As depicted in Fig. 7.4C, when the sandwich immunoassay, which uses detection antibody-conjugated RJPs as the signaling probe, is conducted, the RJPs that are bound to the sensing surface by immunological interaction are observed as shining dots. The observed RJPs were counted using ImageJ software. The number of RJPs was increased along with the concentration of the antigen. The quantitative analysis of cardiac troponin-I (cTnI), a biomarker of acute myocardial infarction, was accomplished with the outstanding limit of detection value as 0.05 ng/mL. In addition, for detecting the signaling probe (RJPs), only a white LED and CMOS camera were used without a sophisticated optical system. Consequently, the sensitive and simple biosensing platform based on the retroreflection phenomenon was developed. Moreover, the retroreflective biosensing platform was used for LAMP-based molecular diagnosis of *Salmonella* and environmental pollutant detection of the mercury ion with a nonspectroscopic optical system [45,46].

In contrast, Willson's group used the microsize corner cube-type retroreflector as a nonspectroscopic ultrabright optical biosensing probe [47]. Using the corner cube-type microretroflector, the immunoassays for *E. coli* and MS2 bacteriophage were designed with detection limits of 10^4 cells/mL and 10^4 virus particles/mL, respectively [48].

The advantage of the retroreflective biosensing platform is the simple optical system for measurement that could be easily imported to the smartphone. The white LED and CMOS sensor were replaced by a smartphone-embedded flashlight and camera module, respectively. As the retroreflective biosensing platform does not require a powerful energy source and complicated detection device, the retroreflective biosensing platform is enough for use as a POCT device that can be used in resource-limited environments. Therefore, the smartphone-based retroreflective biosensing platform with the microfluidic system can be widely used from immunoassay to molecular diagnostics.

2.4 Microscopic bioimaging using the smartphone-embedded camera

Traditionally, the method applied for diagnosing disease is a cell counting method. The marker cell of disease will show an increased or decreased level under disease conditions. To detect the concentration of marker cells, optical microscopy is generally used. The microscopic detection of cells is conducted with cell counting chamber such as hemocytometer in conjunction with cell staining dye. As counting cells with optical microscopy is a time-consuming and labor-intensive process, automatic cell counter was developed. However, the cell counting device is difficult to use in resource-limited conditions due to needs of trained technicians and unreasonable price. Therefore, using the camera of the smartphone for detecting marker cells is a promising tool in the diagnosis of the disease even in resource-limited conditions [49,50]. As an example, Shafiee's group conducted label-free CD4$^+$ T cell detection using the smartphone. In this study, they diagnose HIV infection by tracking the concentration of CD4$^+$ T cells that decrease by HIV infection. For this, a microfluidic device that immobilized the anti-CD4$^+$ T cell antibody was used. When the sample that contained the human CD4$^+$ T cells was injected into the microfluidic device, the CD4$^+$ T cells were captured by the antibody. The captured CD4$^+$ T cells could be observed with a smartphone-embedded optical system that has an additional commercial lens for magnification of the cell image. As a light source for detection, an LED operated by a 3 V battery is required to illuminate the light for the microfluidic channel. Using this simple microfluidic detection system, the Shafiee's group showed that label-free CD4$^+$ T cells could be identified to indicate HIV positive/negative status using a smartphone. The results from detecting the cells using a smartphone were comparable to those obtained using conventional fluorescence-activated cell sorting (FACS) equipment that is used for cell counting and characterization [51]. Similarly, other marker cells present in the biofluid or blood can be measured by using smartphone-based live cell measurement systems that correlate with the FACS device [49,52−54]. In these smartphone-based bioimaging systems, a vital component that is required for cell detection is a light source that exists across the microfluidics. To this end, an additional battery needs to operate the LED. This additional component becomes a drawback in the manufacturing of the POCT device. To eliminate the need for a battery and LED to construct a simple structure that can be used in resource-limited conditions, a platform that can use ambient light as a light source has also been developed [55].

Including diagnostic methods, bioimaging using the smartphone can be used in many areas. For example, blood viscosity measurements, hematocrit measurements, and ABO blood type determination are possible [29,56−58]. In addition, a chemotaxis assay and wound healing assay are possible through observation of cell migration [59,60]. If the fluorescence probe is used with a smartphone-based bioimaging platform, it is also possible to detect a single virus through the smartphone [61]. The more specific description about smartphone-based microscopes is present in Chapter 9.

Smartphone-based bioimaging has been commercialized for the self-check testing of sperm condition. Therefore, smartphone-based bioimaging will contribute to more home-based medical self-checking in the future.

2.5 Other optical detection technologies using the smartphone-embedded camera

As mentioned earlier, the smartphone-based optical detection method can be applied to colorimetry, fluorescence, the retroreflective biosensing platform, and bioimaging. In this section, we will cover other methods that use the smartphone-embedded camera for identifying biomarkers and biochemicals.

First, there is a surface plasmon resonance (SPR) device using a smartphone-embedded camera and LCD screen [62−65]. The smartphone-based SPR device consists of the LCD screen as the light source and a selfie camera that exists in front of the smartphone used as a light detector. Emitted light from the LCD screen of the smartphone is refracted by an optical coupler composed of several prisms and PDMS. The refracted light-illuminated biosensing surface is a capture antibody-immobilized metallic thin film. When an antigen is injected into the sensing surface through the microfluidic system and forms an immunocomplex, a resonance shift of the surface, which causes a change in the incidence angle and a change in reflected light intensity, occurs. As this change correlates with the amount of antigen, the amount of antigen can be detected. Therefore, the SPR-based detection device has the advantage of quantitatively analyzing biomarkers that are present in the biofluid using a label-free immunoassay. However, as SPR devices require an external pump to operate the microfluidic system, a smartphone-based SPR device is not suitable for POCT. In addition, the complicated optimizing process required for the LCD screen and front camera according to each smartphone's specifications is still a disadvantage.

Next, there is a quick response (QR) code-based biosensing platform [66−68]. The detection method as mentioned earlier is used for measuring the brightness of light or imaging the object. For analyzing the obtained image, the smartphone application for analyzing RGB value, the counting algorithm, and other analysis algorithms are required for data analysis. However, developing these new analytical applications in the research laboratory is time consuming. In addition, a lab-made analytical application is difficult for the end user to operate. Therefore, to enable analysis without the help of smartphone analytical applications, analysis of data should be processed by the smartphone without lab-made analytical applications. The QR code can be recognized easily through the camera application embedded in the smartphone without other professional applications. The QR code consists of square grids that form a specific pattern necessary for recognition. If the pattern of this region is represented in a blank, the QR code is not recognized. In contrast, when the specific pattern region is filled with color, which occurs during a biological reaction, the QR code is recognized by the smartphone. Yuan and researchers proposed a QR code-based immunoassay platform. In the proposed platform, when the capture antibody is immobilized on the specific pattern region of the PAD and biofluidics is applied to the PAD using lateral flow, the sandwich-type immunocomplex will be formed in the antibody-immobilized region in the PAD according to the biomarker or biochemical concentration. The GNP-conjugated detection-antibody present in the specific region changes to a black color by silver staining. The QR code that has a specific region filled with black color is recognized by the smartphone. Therefore, the QR code-based detection method can be used as an on–off-type diagnostic platform in a resource-limited condition by an untrained person.

3. Microfluidic-based medical diagnostics using the illumination sensor of the smartphone

In Section 2.1, we described how the antigen and glucose can be quantitatively analyzed by measuring absorbance or reflected light from a sample using the smartphone-embedded camera. Instead of the smartphone-embedded camera, the illumination sensor on the smartphone also can measure the concentration of the sample [69]. This section covers how to measure the concentration of the substance with the illumination sensor of the smartphone.

The illumination sensor consists of a photodiode that converts light into an electrical signal and quantifies the amount of light per unit area. Generally, the brightness of the smartphone screen changes according to the amount of light received from the illumination sensor of the smartphone, making it easier for the user to view the screen of the smartphone even when the ambient light is bright. In addition to these roles, the illumination sensor can also be used as the light receiver of the biosensor. The illumination sensor can convert the received light to a numerical value known as lux. The lux value varies with the concentration of the light-interfering material if the light-interfering material present on the illumination sensor obscures the illumination sensor.

As described in Fig. 7.5A, a chromophore produced by biochemical reactions can cause light interference and reduce the amount of light passing through the illumination sensor. Based on this principle, Park et al. developed an optical biosensor with a smartphone-embedded illumination sensor [70]. In the study by Park et al., the light source and receiver were replaced by the white flashlight existing in the back of the smartphone and the illumination sensor embedded in the smartphone, respectively. Using the smartphone-embedded illumination sensor and flashlight, to combine with the immunoblotting assay, a microfluidic channel consisting of PDMS and PET film was placed between the flashlight and the illumination sensor. In the sensing channel, an insoluble precipitate was formed as a result of the immunoblotting assay. In this study, uCTX-II, which is an osteoarthritis marker, was selected as a target antigen. As depicted in Fig. 7.5B, for a competitive assay for uCTX-II, the polydopamine-coated microfluidic channel was used as a sensing channel because of transparency, the simplicity of fabrication, and immobilization of the biomolecule. After immobilization of PEG$_4$-EKGPDP, an analog of uCTX-II, in the microfluidic, the HRP-conjugated anti-uCTX-II antibody was mixed with the uCTX-II sample and injected into the sensing channel. After washing the unreacted antibody, 3,3′,5,5′-Tetramethylbenzidine membrane (TMBM) was injected into the microfluidic sensing channel. As the concentration of uCTX-II increases, the HRP-conjugated anti-CTX-II antibody which binds to PEG$_4$-EKGPDP

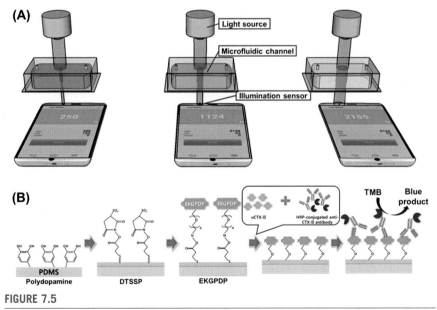

FIGURE 7.5

(A) Schematic illustration of a biosensor using smartphone-embedded illumination sensor proposed by Park et al. (B) Mechanism of a competitive immunoassay that produces light interference material in the microfluidic channel.

immobilized in the channel is decreased. The oxidized TMBM, which is produced by HRP, is also reduced, resulting in a decreased insoluble blue precipitate. As this insoluble precipitate works as a light-interfering agent, the penetration of the LED light was reduced and then the amount of light reaching the sensor was also reduced. Therefore, the amount of light decreases at an inverse proportion to the amount of insoluble precipitate. Consequently, the authors have shown that uCTX-II concentration can be quantitatively analyzed in a microfluidic channel using a flashlight and illumination sensor of the smartphone. However, to illuminate the smartphone flashlight on the illumination sensor, the optical cable is needed to lead the light of the smartphone LED to the illumination sensor. In addition, so that the influence of the external light source is minimized, a dark room condition must be provided for the developed platform, resulting in a large gadget and low portability. Therefore, a study of the simpler biosensor using a smartphone-embedded illumination sensor needs to be conducted.

The Tang's group made a simple instrument with a cylindrical shape (diameter; 14 mm, length; 38 mm) for biosensing with an illumination sensor [71]. This device provides a platform for measuring the concentration of a substance by observing the LSPR peak shifting of the triangular silver nanoprisms (AgNPRs) induced by a target substance. Using the developed platform, the carcinoembryonic antigen (CEA) and adenosine triphosphate (ATP) could be quantitatively analyzed. The device was smaller than the size of a finger, but there were the disadvantages that the battery was inherent and only a target-specific wavelength was used.

To compensate for this, Park et al. developed a novel platform that has the capability of measuring at various light conditions using a smartphone-embedded illumination sensor [72]. In the previous study, the amount of light entering the illumination sensor was kept constant by the provision of the smartphone flashlight and the dark room condition; however, the incidental equipment for dark conditions reduces the portability. To use ambient light instead of a smartphone-embedded flashlight, the ambient light that enters the smartphone illumination sensor should be revised according to time, weather, and location. Therefore, to measure uCTX-II in various ambient light conditions, the concept of transmittance (%, the ratio of reference lux to registered lux) was introduced. Consequently, the authors confirmed that the uCTX-II immunoblotting assay using the smartphone-embedded illumination sensor could use not only an embedded flashlight but also various light conditions such as indoor lighting, direct sunlight, and shade conditions. Using the developed platform, the immunoassay can be conducted regardless of light conditions.

In this way, the biosensor using the illumination sensor can be used as a portable POCT device because the equipment required for diagnosis is only a microfluidic channel that is inexpensive and simple. The untrained end users can efficiently operate the device, and it does not require a high-voltage power supply.

4. Electrochemical detection method using a smartphone integrated with microfluidics

The electrochemical analysis method is widely used for biosensing because of the simplicity of operating and data processing, minimum sample requirement, and low limit-of-detection value [73]. A typical example of an electrochemical analysis device is a handheld-type blood glucose meter. Compared to lab-scale blood glucose analyzers, a handheld-type blood glucose meter has simplicity and portability. However, the inconvenience of carrying the blood glucose meter remains a problem. To increase the portability and convenience for end users, the smartphone could be used as an optimal platform for implantation of the electrochemical analyzer. Because the smartphone is an indispensable item for people, if the smartphone can be used as a glucose detection analyzer, another portable device that includes a heavy battery is not required for analysis of blood glucose detection. In addition to this advantage, for electrochemical measurements, the smartphone has a battery for driving the OLED liquid crystal, applications, and a central processing unit for data processing and electrochemical analysis. Therefore, smartphone-based electrochemical platform can be all-in-one platform for biochemical measurements by using with the enzyme-mediated detection methods [74–77].

As well as biochemical measurements, the immunoassay of a biomarker is possible with a smartphone-embedded electrochemical analysis instrument [78]. Lillehoj and coworkers developed an electrochemical immune-diagnosis biosensor that can be plugged into the micro-USB port of the smartphone. As shown in Fig. 7.6A, the micro-USB cable was mediated by a smartphone and a circuit board which connected to the electrodes present in the microfluidic system. Using the

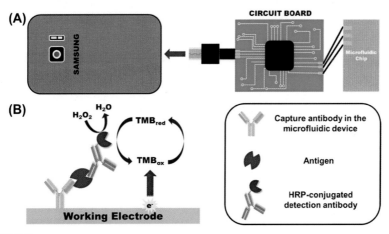

FIGURE 7.6

(A) Schematic illustration of an electrochemical biosensor using smartphone-embedded battery proposed by Lillehoj et al. (B) Mechanism of an electrochemical immunoassay.

microfluidic system, two materials can be simultaneously injected and mixed through a microfluidic mixer driven by capillary force without an external pump. When the mixed sample is injected into the electrodes in the microfluidic chip, electrochemical signaling occurs as shown in Fig. 7.6B. When TMB is introduced into the microfluidic system, HRP, which is conjugated to the detection antibody, is oxidized by TMB and reduced to H_2O_2. The oxidized TMB (TMB_{ox}) receives electrons from the electrode and is changed to the reduced form of TMB (TMB_{red}). Therefore, a decrease in the current value occurs when an amperometric measurement is performed. Using the proposed principle, the quantitative analysis of the biomarker (*Plasmodium falciparum* histidine-rich protein 2 (*Pf*HRP2); biomarker of malaria) in serum by using the smartphone-based electrochemical detection method was accomplished. In addition to the micro-USB port, an audio jack also can be utilized as a useful tool for connecting a smartphone with a microfluidic chip to create a smartphone-based electrochemical sensor. Hall's group and other groups developed an electrochemical sensor using a standard 3.5 mm audio jack of a smartphone [79−83]. In the proposed research, a screen-printed electrode connected to an audio jack was used for sensing surface for measuring biochemicals and biomarkers. Currently, only screen-print electrodes are used as the sensing surface. In the future, if the microfluidic sensing system is combined with the electrode, the smartphone audio jack-based electrochemical biosensing platform is expected to become more widely used and commercialized.

For smartphone-based electrochemical detection methods, electron-transferring mediators are required. Various electron-transferring mediators such as ferrocene and derivatives, tetracyanoquinodimethane (TCNQ), tetrathiafulvalene (TTF), quinones, and ferri/ferrocyanide can be used [84]. Among these, when $[Ru (Bpy)_3]^{2+}$/TPA is used as an electron-transferring mediator, electrochemiluminescence (ECL) can be obtained as a signal in addition to an electric signal [85]. Therefore, the smartphone can be used as a portable ECL device, as the smartphone has a power supply that can operate ECL using the $[Ru (Bpy)_3]^{2+}$ mediator and a camera that can be used as a detector to detect the 620 nm photon, which was generated by $[Ru (Bpy)_3]^{2+}$. The advantages of ECL, including high signal-to-noise ratio, high sensitivity, and precise control of the electrochemical reaction, could be materialized in a smartphone [86,87]. However, the existing studies regarding the smartphone-based ECL device do not use the microfluidic system. Therefore, the advantages of a microfluidic system were not applied to the smartphone-based ECL device. In addition, currently studied smartphone-based portable ECL devices usually do not use the smartphone-embedded battery for ECL generation but use only a smartphone-embedded camera as a detector for luminescence [88−90]. In such cases, accomplishing portability is difficult. Therefore, research on a portable ECL device using the smartphone-embedded battery and microfluidic system should be advanced.

5. Smartphone-embedded data processing applications and the wireless network

As the development of technology, the central processing unit (CPU) of the smartphone has a similar performance to conventional computers with respect to the processing of a massive amount of information and operating applications. Therefore, it is possible to analyze biosensing data in the smartphone as an alternative to using computers such as a desktop or laptop, which were previously used for analyzing data. In this section, we cover the research regarding the use of internal smartphone applications for data analysis.

To analyze the biosensing data using the smartphone, an analysis program such as ImageJ used in the computer should be ported to the smartphone. The porting process was conducted in a researcher's laboratory, and the lab-made analysis application was used for analyzing biosensing data. Most of the papers mentioned earlier adopted lab-made applications for biosensing data processing [20,21,26,31,33,39,49,52,72,74,76,78−82]. When the smartphone-based internal application for biosensing data processing was used, it was possible to analyze the biosensing data immediately on the smartphone without an additional process. Thus, the smartphone internal data analysis application is more suitable as a POCT device. In addition, with the development of the CPU, more complex analysis programs and algorithms such as deep learning will be possible using a smartphone.

Conversely, a case occurred in which only a smartphone-embedded application was used for diagnosis instead of other smartphone-embedded components. In this case, the biosensing data that were acquired from benchtop-size detection devices were transferred to the smartphone using wireless communication and analyzed by a smartphone-embedded application. The external diagnostic device was utilized only to obtain biosensing data from the sample by using the biosensing technique such as colorimetric assay, fluorescence assay detection, or electrochemical analysis. Then, a smartphone-embedded application analyzed the transmitted data via the wireless communication system such as Bluetooth and Wi-Fi [91−97]. In this case, although user convenience could be increased, it is difficult to use as a POCT device because the smartphone was not used for biosensing as stand-alone.

When the data are transmitted to the web server using the wireless communication technology such as WiFi and cellular data (e.g., LTE) and the data are automatically uploaded through the internet, monitoring of diagnostic data by professional doctors is achieved and professional feedback will also be available. Moreover, patient history data are stored so that it is easy to observe the disease continuously. In addition, as the smartphone can obtain the position information using a GPS, tracking of patients, mapping of diseases, and statistical analysis of big data will be possible. From now on, the development of cloud computing will enable personalized medicine [98−100].

6. The diagnostic methods using smartphone-embedded NFC technology

The smartphone-based data processing platform needs wireless communication technology, which is capable of transferring and receiving data. Short-range wireless communication technologies including Bluetooth, WiFi, and NFC, which are implanted in the smartphone, are used for transferring the diagnostic data between the smartphone-based device and external device. Among these wireless communication tools, NFC can be used for simple data transmission with short-range wireless communication. In addition, NFC itself is used as a platform for the biosensor [101−104]. Rose et al. proposed an electrochemical sweat ion analysis platform using the microfluidic system and electrodes connected to the RFID tag, which could be read by a smartphone-embedded NFC sensor [105]. The smartphone-embedded NFC sensor provides electromagnetic waves to the RFID tag. The RFID tag converts electromagnetic waves to electrical energy using energy harvesting circuitry. Using electrical energy, electrochemical analysis of the sweat ion can be performed on the electrodes connected to the RFID. Depending on the ion concentration in the sweat, the potential of the electrode increases. The changed potential value can be read by the smartphone-embedded NFC sensor through the antenna of the RFID tag. Therefore, the concentration of the sweat ion can be measured by the smartphone-embedded NFC sensor. In addition to sweat ion measurement, if an enzyme such as uricase is immobilized on this electrode, uric acid in the sweat can also be measured [106]. Hence, various biochemicals can be measured using the enzyme-mediated detection method.

In addition, the immunoassay is also possible using the smartphone-based NFC reader [107]. According to the principle proposed by Yuan et al., in a disconnected RFID tag, the immunoassay was performed in the NC membrane, which was present in the vacant region of the RFID tag. When the immunological reaction was accomplished with the GNP-conjugated detection antibody and silver staining method, the RFID tag was connected and the signal turned on. The immunologically induced RFID signal was read by the smartphone-embedded NFC sensor. Therefore, the smartphone NFC-based signal on−off-type biosensing platform was achieved. Using the smartphone-embedded NFC sensor, the battery-free operation was also achieved by energy harvesting circuitry. Combining these tools, the use of medical diagnostics in resource-limited conditions is expected.

7. Conclusions

We have witnessed significant advances in microfluidics technology and microfluidic devices being used in biomedical applications. Now, this technology has found an important contact point with handset devices (smartphones), especially in view of biosensing applications. Sensor elements, data acquisition and processing, and communication devices, all built into smartphones, could realize the marriage between microfluidic devices and smartphones. Much progress has been made in the field where built-in equipment is employed in optical and electrochemical analyzers for biosensing, data processing, or data communications. The resulting stand-alone microfluidic-biosensing device is user friendly and small in size, making it ideal for use in field diagnosis (POCT). Based on the tendency of increased reporting of research results in related fields at academic conferences and journals, practical applications of these technologies will be found in many fields soon, leading to actual commercialization.

Acknowledgments

This study was supported by grants from Creative Materials Discovery Program (2019M3D1A1078938) and Priority Research Center Program (2019R1A6A1A11051471) funded by the National Research Foundation of Korea (NRF).

References

[1] Wentao S, Xinghua G, Lei J, Jianhua Q. Microfluidic platform towards point-of-care diagnostics in infectious diseases. J Chromatogr A 2015;1377:13–26.

[2] Catherine R, Hyewon I, Alison H, et al. Microfluidics for medical diagnostics and biosensors. Chem Eng Sci 2011;66:1490–507.

[3] Bambang K, Nuriman, Jurriaan H, et al. Optical sensing systems for microfluidic devices: a review. Anal Chim Acta 2007;601:141–55.

[4] Daniel QG, Arben M. Mobile phone-based biosensing: an emerging "diagnostic and communication" technology. Biosens Bioelectron 2017;92:549–62.

[5] Lenka H, Jan P. Smartphones & microfluidics: marriage for the future. Electrophoresis 2018;39:1319–28.

[6] Mohammad Z. Portable biosensing devices for point-of-care diagnostics: recent developments and applications. Trends Anal Chem 2017;91:26–41.

[7] Francesco L, Giuseppe P, Kurt B, et al. Health parameters monitoring by smartphone for quality of life improvement. Measurement 2015;73:82–94.

[8] Pasquale D, Luca DV, Francesco P, et al. State of the art and future developments of measurement applications on smartphones. Measurement 2013;46:3291–307.

[9] Suvardhan K, Myalowenkosi IS, Phumlane SM, et al. Smartphone based bioanalytical and diagnosis applications: a review. Biosens Bioelectron 2018;102:136–49.

[10] Kenneth DL, Hojeong Y, Brian TC. Smartphone instrument for portable enzyme-linked immunosorbent assays. Biomed Opt Express 2014;5(11):3792–806.

[11] Ahmet FC, Justin W, Delaram K. A personalized food allergen testing platform on a cellphone. Lab Chip 2013;13:636.

[12] Hyeong JC, Yong DH, Yoo MP. An optical biosensing strategy based on selective light absorption and wavelength filtering from chromogenic reaction. Materials 2018;11(3):388.

[13] Jessica MDM, Ruben RGS, Virginia C, et al. Multiplexed capillary microfluidic immunoassay with smartphone data acquisition for parallel mycotoxin detection. Biosens Bioelectron 2018;99:40–6.

[14] Heba AA, Hassan MEA. Power-free chip enzyme immunoassay for detection of prostate specific antigen (PSA) in serum. Biosens Bioelectron 2013;49:478—84.

[15] Kangning R, Jianhua Z, Hongkai W. Materials for microfluidic chip fabrication. Acc Chem Res 2013;46(11):2396—406.

[16] Ho NC, Yiwei S, Bin X, et al. Simple, cost-effective 3D printed microfluidic components for disposable, point-of-care colorimetric analysis. ACS Sens 2016;1(3): 227—34.

[17] Chanyong P, Yong DH, Han VK, et al. Double-sided 3D printing on paper towards mass production of three-dimensional paper-based microfluidic analytical devices (3D-μPADs). Lab Chip 2018;18:1533—8.

[18] Yoshiaki U, Yuichi U, Yuzuru T. Direct digital manufacturing of a mini-centrifuge-driven centrifugal microfluidic device and demonstration of a smartphone-based colorimetric enzyme-linked immunosorbent assay. Anal Methods 2016;8:256.

[19] Kimberly P, Matthew C, Timothy M, et al. 3D printed auto-mixing chip enables rapid smartphone diagnosis of anemia. Biomicrofluidics 2016;10:054113.

[20] Andres WM, Scott TP, Emanuel C, et al. Simple telemedicine for developing regions: camera phones and paper-based microfluidic devices for real-time, off-site diagnosis. Anal Chem 2008;80:3699—707.

[21] Nuria LR, Vincenzo FC, Miguel ME, et al. Smartphone-based simultaneous pH and nitrite colorimetric determination for paper microfluidic devices. Anal Chem 2014; 86:9554—62.

[22] Yanyan X, Jin S, Zhiyang L. Fabrication techniques for microfluidic paper-based analytical devices and their applications for biological testing: a review. Biosens Bioelectron 2016;77:774—89.

[23] Sangdae L, Giyoung K, Jihea M. Performance improvement of the one-dot lateral flow immunoassay for aflatoxin B1 by using a smartphone-based reading system. Sensors 2013;13:5109—16.

[24] Raphael CW, Harley YT. Lateral flow immunoassay. Humana Press, Springer; 2009.

[25] Evgeni E, Sarah G, Adarina LYK, et al. Lateral flow immunoassays — from paper strip to smartphone technology. Electroanalysis 2015;27:2116—30.

[26] Meili D, Jiandong W, Zimin M, et al. Rapid and low-cost CRP measurement by integrating a paper-based microfluidic immunoassay with smartphone (CRP-chip). Sensors 2017;17(4):684.

[27] Hyeong JC, Yoo MP, Yong DH, et al. Paper-based glucose biosensing system utilizing a smartphone as a signal reader. Biochip J 2014;8(3):218—26.

[28] Seong HI, Ka RK, Yoo MP, et al. An animal cell culture monitoring system using a smartphone-mountable paper-based analytical device. Sens Actuators B Chem 2016; 229:166—73.

[29] Gang C, Hui HC, Ling Y, et al. Smartphone supported backlight illumination and image acquisition for microfluidic-based point-of-care testing. Biomed Opt Express 2018;9(10):4604—12.

[30] Li S, Joshua AH, Ian P. Point-of-care colorimetric detection with a smartphone. Lab Chip 2012;12:4240—3.

[31] Ahyeon K, Daeshik K, Yeguang X, et al. A soft, wearable microfluidic device for the capture, storage, and colorimetric sensing of sweat. Sci Transl Med 2016;8. 366ra165.

[32] Ana IB, Poonam G, Kalpita S, et al. Portable smartphone quantitation of prostate specific antigen (PSA) in a fluoropolymer microfluidic device. Biosens Bioelectron 2015; 70:5—14.

[33] Ahmet FC, Justin W, Delaram K, et al. A personalized food allergen testing platform on a cellphone. Lab Chip 2013;13:636—40.

[34] Kenneth DL, Elizabeth VW, Huy ML, et al. Multimode smartphone biosensing: the transmission, reflection, and intensity spectral (TRI)-analyzer. Lab Chip 2017;17: 3246—57.

[35] Vinoth KR, Padmavathy B, Baquir MJA. Smartphone based bacterial detection using biofunctionalized fluorescent nanoparticles. Microchim Acta 2014;181:1815—21.

[36] Chenji Z, Jimin PK, Michael C, et al. A smartphone-based chloridometer for point-of-care diagnostics of cystic fibrosis. Biosens Bioelectron 2017;97:164—8.

[37] Ali KY, Nan J, Ali T, et al. Paper-based microfluidic system for tear electrolyte analysis. Lab Chip 2017;17:1137—48.

[38] Yurina S, Sung BK, Yi Z, et al. A fluorometric skin-interfaced microfluidic device and smartphone imaging module for in situ quantitative analysis of sweat chemistry. Lab Chip 2018;18:2178—86.

[39] Ryan S, Andrea G, Li J, et al. KS-Detect—Validation of solar thermal PCR for the diagnosis of Kaposi's sarcoma using pseudo-biopsy samples. PLoS One 2016;11(1): e0147636.

[40] Jinzhao S, Changchun L, Michael GM, et al. Two-stage isothermal enzymatic amplification for concurrent multiplex molecular detection. Clin Chem 2017;63(3):714—22.

[41] Weili C, Hojeong Y, Fu S, et al. Mobile platform for multiplexed detection and differentiation of disease-specific nucleic acid sequences, using microfluidic loop-mediated isothermal amplification and smartphone detection. Anal Chem 2017;89(21): 11219—26.

[42] Shih-Chuan L, Jing P, Michael GM, et al. Smart cup: a minimally-instrumented, smartphone-based point-of-care molecular diagnostic device. Sens Actuators B Chem 2016;229:232—8.

[43] Mark HB, Jacqueline N, Christopher MC, et al. Retroreflective imaging system for optical labeling and detection of microorganisms. Appl Opt 2014;53:3647—55.

[44] Yong DH, Hyo-Sop K, Yoo MP, et al. Retroreflective Janus microparticle as a nonspectroscopic optical immunosensing probe. ACS Appl Mater Interfaces 2016;8(17): 10767—74.

[45] Hyeong JC, Saemi K, Yong DH, et al. Water-soluble mercury ion sensing based on the thymine-Hg^{2+}-thymine base pair using retroreflective Janus particle as an optical signaling probe. Biosens Bioelectron 2018;104:138—44.

[46] Hyeong JC, Seongok K, Yong DH, et al. *Salmonella* Typhimurium sensing strategy based on the loop-mediated isothermal amplification using retroreflective Janus particle as a nonspectroscopic signaling probe. ACS Sens 2018;3(11):2261—8.

[47] Tim S, Azeem N, Julia L, et al. Suspended, micron-scale corner cube retroreflectors as ultra-bright optical labels. J Vac Sci Technol B 2011;29:06FA01.

[48] Gavin G, David S, Federico RR, et al. Microretroreflector-sedimentation immunoassays for pathogen detection. Anal Chem 2014;86(18):9029—35.

[49] Hongying Z, Ikbal S, Justin W, et al. Cost-effective and rapid blood analysis on a cellphone. Lab Chip 2013;13(7):1282—8.

[50] Derek T, Onur M, Cetin O, et al. Lensfree microscopy on a cellphone. Lab Chip 2010; 10:1787—92.

[51] Manoj KK, Hardik JP, Mohamed SD, et al. Rapid, label-free CD4 testing using a smartphone compatible device. Lab Chip 2017;17:2910—9.

[52] Yujin Z, Ke J, Jie L, et al. A low cost and portable smartphone microscopic device for cell counting. Sens Actuators A Phys 2018;274:57−63.

[53] Manoj KK, Magesh S, Anupriya S, et al. An automated smatphone-based diagnostic assay for point-of-care semen analysis. Sci Transl Med 2017;9:eaai7863.

[54] Hongying Z, Sam M, Ahmet FC, et al. Optofluidic fluorescent imaging cytometry on a cell phone. Anal Chem 2011;83:6641−7.

[55] Seung AL, Changhuei Y. A smartphone-based chip-scale microscope using ambient illumination. Lab Chip 2014;14:3056−63.

[56] Sang CK, Uddin MJ, Sung BI, et al. A smartphone-based optical platform for colorimetric analysis of microfluidic device. Sens Actuators B Chem 2017;239:52−9.

[57] Uddin MJ, Sang CK, Joon SS. Histogram analysis for smarphone-based rapid hematocrit determination. Biomed Opt Express 2017;8(7):3317−28.

[58] Sooyeong K, Kyung CK, Eunseop Y. Microfluidic method for measuring viscosity using images from smartphone. Opt Lasers Eng 2018;104:237−43.

[59] Ke Y, Jiandong W, Hagit PS, et al. M_{kit}: a cell migration assay based on microfluidic device and smartphone. Biosens Bioelectron 2018;99:259−67.

[60] Rodrigo HV, Emil S, Nikos FK, et al. A modular and affordable time-lapse imaging and incubation system based on 3D-printed parts, a smartphone, and off-the-shelf electronics. PLoS One 2016;11(12):e0167583.

[61] Qingshan W, Hangfei Q, Wei L, et al. Fluorescent imaging of single nanoparticles and viruses on a smartphone. ACS Nano 2013;7(10):9147−55.

[62] Pakorn P, Marcos CG, Anke S, et al. Surface plasmon resonance chemical sensing on cell phones. Angew. Chemie. Weinheim. Bergstr. Ger. 2012;124(46):11753−6.

[63] Hasan G, Erol O, Guzin K, et al. A smartphone based surface plasmon resonance imaging (SPRi) platform for on-site biodetection. Sens Actuators B Chem 2017;239:571−7.

[64] Carlos ASF, Antonio MNL, Helmut N. Smarphone based, portable optical biosensor utilizing surface plasmon resonance. IEEE Conf Int Instru Meas Tech. 2014:1091−5281.

[65] Yun L, Qiang L, Shimeng C, et al. Surface plasmon resonance biosensor based on smart phone platforms. Sci Rep 2015;5:12864.

[66] Mingquan Y, Keng-ku L, Srikanth S, et al. Self-powered forward error-correcting biosensor based on integration of paper-based microfluidics and self-assembled quick response codes. IEEE Trans Biomed Circuits Syst 2016;10(5):963−71.

[67] Tengfei W, Guangning X, Wenyuan W, et al. A novel combination of quick response code and microfluidic paper-based analytical devices for rapid and quantitative detection. Biomed Microdevices 2018;20:79.

[68] Alison B, Harrison KSH, Flavio AF, et al. Printable QR code paper microfluidic colorimetric assay for screening volatile biomarkers. Biosens Bioelectron 2019;128:97−103.

[69] Sibasish D. Point of care sensing and biosensing using ambient light sensor of smartphone: critical review. Trends Anal Chem 2019;110:393−400.

[70] Yoo MP, Yong DH, Ka RK, et al. An immunoblot-based optical biosensor for screening of osteoarthritis using a smartphone-embedded illuminometer. Anal Methods 2015;7:6137−42.

[71] Qiangqiang F, Ze W, Fangxiang X, et al. A portable smart phone-based plasmonic nanosensor readout platform that measures transmitted light intensities of nanosubstrates using an ambient light sensor. Lab Chip 2016;16:1927−33.

[72] Yoo MP, Yong DH, Hyeong JC, et al. Ambient light-based optical biosensing platform with smartphone-embedded illumination sensor. Biosens Bioelectron 2017;93: 205—11.

[73] Xuefei G, Nianqiang W. Smartphone-based sensors. Electrochem Soc Interface 2016; 25(4):79—81.

[74] Jinhong G. Smartphone-powered electrochemical dongle for point-of-care monitoring of blood β-ketone. Anal Chem 2017;89(17):8609—13.

[75] Jinhong G. Uric acid monitoring with a smartphone as the electrochemical analyzer. Anal Chem 2016;88(24):11986—9.

[76] Joan A, Augusto M, Lluís T, et al. Cost-effective smartphone-based reconfigurable electrochemical instrument for alcohol determination in whole blood samples. Biosens Bioelectron 2018;117:736—42.

[77] Dandan X, Xiwei H, Jinhong G, et al. Automatic smartphone-based microfluidic biosensor system at the point of care. Biosens Bioelectron 2018;110:78—88.

[78] Peter BL, Ming-Chun H, Newton T, et al. Rapid electrochemical detection on a mobile phone. Lab Chip 2013;13:2950.

[79] Alexander S, Travis W, Akshay GV, et al. A low-cost smartphone-based electrochemical biosensor for point-of-care diagnostics. IEEE Biomed Circuits Syst Conf 2014: 312—5.

[80] Alexander CS, Chengyang Y, Akshay GV, et al. An efficient power harvesting mobile phone-based electrochemical biosensor for point-of-care health monitoring. Sens Actuators B Chem 2016;235:126—35.

[81] Eliah AS, Akshay GV, Alex S, et al. Detection of Hepatitis C core antibody by dual-affinity yeast chimera and smartphone-based electrochemical sensing. Biosens Bioelectron 2016;86:690—6.

[82] Xinhao W, Manas RG, Jing J, et al. Audio jack based miniaturized mobile phone electrochemical sensing platform. Sens Actuators B Chem 2015;209:677—85.

[83] Jacqui LD, Egan HD, Anthony JH, et al. Reprint of: use of a mobile phone for potentiostatic control with low cost paper-based microfluidic sensors. Anal Chim Acta 2013; 790:56—60.

[84] Asha C, Bansi DM. Mediated biosensors. Biosens Bioelectron 2002;17:441—56.

[85] Gary FB, Haresh PS, John HK, et al. Electrochemiluminescence detection for development of immunoassays and DNA probe assays for clinical diagnostics. Clin Chem 1991;37(9):1534—9.

[86] Egan HD, Gregory JB, Anthony JH, et al. Mobile phone-based electrochemiluminescence sensing exploiting the 'USB On-The-Go' protocol. Sens Actuators B Chem 2015;216:608—13.

[87] Shuang L, Danhua Z, Jinglong L, et al. Electrochemiluminescence on smartphone with silica nanopores membrane modified electrodes for nitroaromatic explosives detection. Biosens Bioelectron 2019;129:284—91.

[88] Jacqui LD, Conor FH, Junfei T, et al. Electrogenerated chemiluminescence detection in paper-based microfluidic sensors. Anal Chem 2011;83(4):1300—6.

[89] Yong Y, Huijie L, Dan W, et al. An electrochemiluminescence cloth-based biosensor with smartphone-based imaging for detection of lactate in saliva. Analyst 2017;142: 3715.

[90] Lu C, Chunsun Z, Da X. Paper-based bipolar electrode-electrochemiluminescence (BPE-ECL) device with battery energy supply and smartphone read-out: a handheld

ECL system for biochemical analysis at the point-of-care level. Sens Actuators B Chem 2016;237:308−17.

[91] Ruey-Jen Y, Chin-Chung T, Wei-Jhong J, et al. Integrated microfluidic paper-based system for determination of whole blood albumin. Sens Actuators B Chem 2018; 273:1091−7.

[92] John MCCG, Lemmuel LT, Chan-Chiung L, et al. Rapid microfluidic paper-based platform for low concentration formaldehyde detection. Sens Actuators B Chem 2018; 255:3623−9.

[93] Zhaoxiong D, Dongying Z, Guanghui W, et al. An in-line spectrophotometer on a centrifugal microfluidic platform for real-time protein determination and calibration. Lab Chip 2016;16:3604.

[94] Chan-Chiung L, Yao-Nan W, Lung-Ming F, et al. Microfluidic paper-based chip platform for benzoic acid detection in food. Food Chem 2018;249:162−7.

[95] Abkar S, Fatimah I, Shah MU, et al. A microdevice for rapid, monoplex and colorimetric detection of foodborne pathogens using a centrifugal microfluidic platform. Biosens Bioelectron 2018;100:96−104.

[96] Yan F, Juntao L, Yang W, et al. A wireless point-of-care testing system for the detection of neuron-specific enolase with microfluidic paper-based analytical devices. Biosens Bioelectron 2017;95:60−6.

[97] Xinhao W, Guohong L, Guangzhe C, et al. White blood cell counting on smartphone paper electrochemical sensor. Biosens Bioelectron 2017;90:549−57.

[98] Jie H, Xingye C, Yan G, et al. Portable microfluidic and smartphone-based devices for monitoring of cardiovascular diseases at the point of care. Biotechnol Adv 2016;34: 305−20.

[99] Sandeep KV, Marion S. E, John HTL. Commercial smartphone-based devices and smart applications for personalized healthcare monitoring and management. Diagnostics 2014;4(3):104−28.

[100] Moeen H, Alex P, Tolga S, et al. Health monitoring and management using internet-of-things (IoT) sensing with cloud-based processing: opportunities and challenges. IEEE Conf Collab Internet Comput 2015:285−92.

[101] Lindsey KF, James C, Ning Y. Electrochemical detection of biogenic amines during food spoilage using an integrated sensing RFID tag. Sens Actuators B Chem 2014; 202:1298−304.

[102] Joseph MA, Katherine AM, Jens BR, et al. Wireless gas detection with a smartphone via rf communication. Proc Natl Acad Sci USA 2014;111(51):18162−6.

[103] Lindsey KF, Ning Y. RFID tags for wireless electrochemical detection of volatile chemicals. Sens Actuators B Chem 2013;186:817−23.

[104] Nizar L, Wassim B, Rigoberto B, et al. Self-powered piezo-floating-gate smart-gauges based on quasi-static mechanical energy concentrators and triggers. IEEE Sens J 2015; 15(2):676−83.

[105] Daniel PR, Michael ER, Daniel KG, et al. Adhesive RFID sensor patch for monitoring of sweat electrolytes. IEEE Trans Biomed Eng 2015;62(6):1457−65.

[106] Petar K, Jayoung K, Rajan K, et al. Smart bandage with wireless connectivity for uric acid biosensing as an indicator of wound status. Electrochem Commun 2015;56:6−10.

[107] Mingquan Y, Evangelyn CA, Shantanu C. A novel biosensor based on silver-enhanced self-assembled radio-frequency antennas. IEEE Sens J 2014;14(4):941−2.

Digital health for monitoring and managing hard-to-heal wounds

Bijan Najafi, PhD, MSc

*Professor of Surgery, Interdisciplinary Consortium for Advanced Motion Performance (iCAMP),
Division of Vascular Surgery and Endovascular Therapy, Michael E. DeBakey Department of
Surgery, Baylor College of Medicine, Houston, TX, United States*

1. Introduction

Despite many recent advances in wound healing management [1], many wound care specialists will encounter wounds that are "hard to heal," where healing is prolonged over months or never achieved and led to substantial cost to healthcare system, thereby devastating the life of patients and their families. Fortunately, we live in a world where digital technologies are increasingly being integrated into every aspect of our lives, representing an opportunity for creative solutions to prevent or better manage this devastating condition. However, the translation of these technologies toward the clinic is still in its infancy. This chapter will first overview examples of hard-to-heal wounds and will discuss on current challenges and opportunities in improving the management of these hard-to-heal wounds. Then, it will overview current technologies based on digital health, wearables, and mobile health (mHealth) applications, which are shown to be promising to improve management and prevention of hard-to-heal wounds.

2. Major types of hard-to-heal wounds: challenges and opportunities

2.1 Diabetic foot ulcers

An example of a hard-to-heal wound is diabetic foot ulcer (DFU), which is an open sore in the foot and frequently serves as a portal for infection (Fig. 8.1). Failure to heal DFUs is a leading cause of global hospitalization, amputation, disability, and death among people with diabetes [2–5]. The lifetime incidence of DFUs has been estimated to be between 19% and 34% among people with diabetes [2]. Globally, it is estimated that 20 million people currently have an active DFU with a further 130 million either having a history of DFU or the precursor risk factor of diabetic peripheral neuropathy (DPN) and expected to develop a DFU in their lifetime without intervention [2,3]. Diabetic foot care costs represent the single largest

Smartphone Based Medical Diagnostics. https://doi.org/10.1016/B978-0-12-817044-1.00008-9

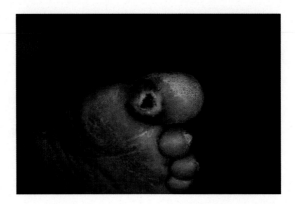

FIGURE 8.1

An example of diabetic foot ulcer (DFU). Excessive stress on plantar tissue over time is one of the leading causes of diabetic foot ulcers among people with diabetes and loss of protective sensation. Because of loss of plantar sensation, of "the gift of pain", from diabetic peripheral neuropathy, people with DFU may not even notice the presence of DFU at early stage leading to worsening the wounds and infection. Healing DFU is complicated because of limited blood flow in people with diabetes and poor adherence to protected footwear. Unfortunately 10%–20% of DFUs end up with amputation. These ulcers never healed, as it is estimated 65% of them reoccurred within 5 years. Recently, wound healing specialists called DFU as a "foot cancer" and a healed DFU as a foot in remission. The good news is that it is estimated that at least 70% of foot amputations because of DFUs are potentially preventable. This opens an opportunity for technology to assist in effective management of DFUs and limits its devastating consequences.

category of excess medical costs associated with diabetes. It is estimated that one-third of all diabetes related costs are spent on diabetic foot care in the United States, with two-thirds of these costs incurred in the inpatient settings, constituting a substantial cost to society [6,7].

Ulcers requiring acute care can result in treatment costs of up to US$28,000 per event, varying with the severity of the wound [8]. Unfortunately, even after the resolution of a foot ulcer, recurrence is common and is estimated to be 40% within 1 year after ulcer healing, almost 60% within 3 years, and 65% within 5 years [2]. Perhaps no subsequent complication of DFU is more significant than its associated 10%–20% rate of lower extremity amputation (LEA) per event. At least 70% of such amputations are potentially preventable [9]. The consequences of DFU are not limited to amputation. In particular, DFUs may put people at risk for other adverse events such as falls, fractures, reduced mobility, frailty, and mortality [10–12]. For example, mortality after amputation because of DFU is estimated to be 70% at 5 years, which exceeds many common cancers such as breast cancer and prostate cancer [13].

Several factors contribute to this high DFU incident and reulceration rate. Frequently, patients with diabetes cannot produce a local inflammatory response that is noticeable to the patient, caregivers, or providers. Providers frequently reappoint patients using their clinical intuition of the patient's average build-up of their preulcerative callus from activity and footwear. However, patients also demonstrate a large variability in activity level that precedes ulceration [1,5,14]. Daily thermometry and plantar stress measures of a patient's feet addresses these fluctuations in activity level and footwear [1,14]. It can provide more objective information for appointment rescheduling [15]; it can help clinicians to determine whether there is an urgent clinical need for a rescheduling even when meaningful change in patient clinical or health status do not warrant one [1,5].

Another major challenge in managing DFUs is poor adherence in wearing offloading (modalities, which are prescribed for relieving high plantar stress under inflamed regions) intervention (e.g., footwear, walkers, casts, etc.). Those with active or history of DFUs as well as those people with impaired biomechanics of lower extremities because of diabetes [14] often require to wear a prescribed footwear including insoles, diabetic shoes, or boots at all time and in particular during weight-bearing activities (e.g., walking and standing). Such prescribed footwear is called "offloading" assist in reducing mechanical stress under plantar regions of interest and thus could protect foot while promoting weight-bearing ambulation. However, these offloading can be only effective when they are worn [16,17]. Thus, any technology that could assist in promoting adherence to prescribed protected footwear could significantly improve management of DFUs.

Efforts are also required to empower patients and their caregivers to be part of health ecosystem including preventing DFUs or taking care of their active DFUs. Recent advances in technology and in particular telecommunication now could facilitate provision of comprehensive and easy to execute feedback (e.g., gamification to engage patients and improve adherence), recommendation (e.g., personalized and easy-to-understand guidelines to manage DFUs), and notification (e.g., alerting about signs of inflammation, which precede skin breakdown) can assist in engaging patients and their caregivers to prevent or improve management of DFUs. For example, everyday measurement and timely feedback of inflammatory response (e.g., change in plantar temperature as a sign of inflammation), high plantar stress (e.g., change in plantar pressure as signs of formation of callus, presence of a foreign object inside of shoes, infective offloading, or altered lower extremities biomechanics because of aging and/or diabetes), and daily activities (e.g., prolonged unbroken walking and standing) have been shown to be effective in improving management of DFUs. However, effective technology is still missing to facilitate these measurements on daily basis, engage patients and their caregivers to use these technologies, as well as personalizing the feedback, simplify alerting system, and effectively communicate with patients, their caregivers, and their care providers. Mobile health (mHealth) technologies could fill the gaps and speed up translation of these effective interventions to become clinical standards for management of DFUs.

2.2 Pressure ulcer

Another example of a hard-to-heal wound is a pressure ulcer (PU), known as bedsore. PUs form on bony prominences (e.g., heels, ankles, hips, and tailbone), usually in cases where people are immobilized for extended periods of time (e.g., bedbound patients, those with spinal cord injuries, etc.), which result in prolonged pressure on the skin. Contributory mechanical forces to PU development include pressure and friction (shear) at various levels and magnitudes. PUs are associated with substantial humanistic, financial, and sociological consequences. Once present, healing is compromised in patients with related comorbidities and places the patient at risk for decreased quality of life, prolonged therapy, sepsis, amputation, and death. It is estimated that 1.3 million to 3 million adults in the United States have a PU, with an estimated cost of $500 to $40,000 to heal each ulcer [18]. Overall, PU prevalence ranges from 7.5% to 18% and incidence ranges from 0.4% to 38% in the United States acute care setting [18,19]. Significant rates have also been found in home, rehabilitative, and long-term care. Critical care patients are at particular high risk where prevalence has been reported to be as high as 29%. It is estimated that up to 50% of PU incidents are occurred at the heel region (Fig. 8.2) [20].

Most PUs are considered preventable. Prevention strategies include risk stratification, prevention protocols, optimal medical management, pressure reduction, and patient and caregiver education. Healthcare regulating bodies require regular risk assessment for PU development with intervention plans based on the assessed risk. Current PU risk assessment strategies have been well studied and have advanced the field. However, risk assessment is not absolute. A large study done in the Netherlands found that 37% of patients who developed ulcerations were not assessed to be at risk for ulcer development [20], indicating the need for better tool to determine true risk. Moreover, even those assessed to be at high risk still developed ulcers despite preventive measures, highlighting an important gap in effective management of PUs.

Current risk assessment offers poor clinical discrimination. A systematic review [21] suggests that "Nursing Intuition" (nurses' clinical judgement) for assessing PU risk has a sensitivity of 51% and specificity of 60%. The most widely used Braden Scale has a sensitivity of 57% and specificity of 68% [21]. Table 8.1 describes the predictive value of these tools using Bayes' theorem with an underlying prevalence of 15%. These results may partially explain why Pancorbo-Hildago et al. [21] concluded that the use of risk assessment scales improves PU preventive interventions, but have not been efficacious in decreasing PU incidence. The success of risk assessment to guide preventive behaviors depends on actual use of the scale. Multiple levels of care have reported struggle with staff adherence to hospital policy regarding use of the tools, despite electronic medical records and clinical templates. Despite implementation of well-studied preventive strategies, PUs continue to occur. This is evident by the national acute-care incidence and prevalence surveys in the United States from 1999 to 2002 that found prevalence never dropped below 14% and incidence was never below 7% [22]. Another recent systematic review [23]

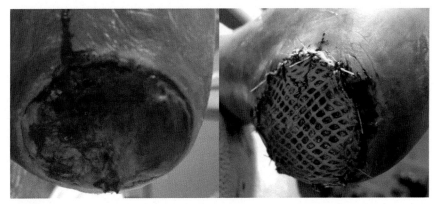

FIGURE 8.2

An example of pressure ulcer (PU) in the heel region. PUs are caused by prolonged pressure because of immobility for extended period of time, and often occurred in older adults. Once PU is present, its management requires advanced and prolonged therapies, and is associated with impressive humanistic, financial, and sociological consequences. The most effective management of PUs is preventing them in the first place. Prevention strategies include risk stratification, prevention protocols, optimal medical management, pressure reduction, and patient/family/caregiver education. mHealth could play an important role from risk stratification to early identify markers of PUs to effectively educate patients and their caregivers in preventing this devastating condition and/or limiting its worsening or consequences.

Table 8.1 Bayes' theorem probability of PU after using intuition or Braden Scale.

Tools	Likelihood ratio	Probability of PU
Nursing intuition		
At risk PU	Likelihood ratio +	23%
Low risk PU	Likelihood ratio −	14%
Braden scale		
At risk PU	Likelihood ratio +	31%
Low risk PU	Likelihood ratio −	11%

that tracked the articles published from 2000 to 2015 also suggests that worldwide PU prevalence rate likely remained high in acute setting and ranged between 6% and 18.5% [23]. Some facilities cite lack of nursing staff or high turnover and lack of administrative commitment as factors suggesting that PUs are not as avoidable as once thought.

Previous investigations have also not elucidated a clear temporal relationship for pressure ulcer onset and development. Gefen [24] reviewed human, animal, and in vitro studies regarding PU timelines that were published from 1966 to 2008. The evidence suggested that PUs in subdermal tissues under bony prominences very likely occur approximately 1—6 h after sustained loading. Pressure ulcer staging systems identify Stage I ulcerations as intact skin with nonblanchable redness of a localized area usually over a bony prominence. However, even in these "preulcers," the evidence suggests that deep tissue injury may already exist [24]. Studies have demonstrated persistent elevations in skin temperatures at the ischial tuberosities of spinal cord injury patients after 3 h of sitting even after 1 h of rest.

Despite resolution of surface erythema, tissue damage may continue to occur even after pressure is reduced [25]. Edsberg reviewed PU studies of tissue histology [26]. Based on the evidence, the author suggested changes visible at the surface of the tissue are minor compared to the damage seen at the deepest layers of tissue. Hence, by the time persistent skin changes are visible, it is often already too late to prevent actual tissue damage. These findings may be more severe in high-risk populations for PU [27].

In summary, the current challenges in effective prevention of the PU is identifying truly high-risk individuals for PU (increase in both sensitivity and specificity), developing sensitive markers that alert caretakers and patients to possible insult, and identifying the most predictive variables for PU development. Technologies may be able to address earlier challenges for effective risk stratification and preventive strategies. A better tool is also needed to capture PU before visible skin damage as by the time PUs are found, significant tissue damage has often already occurred. Few promising efforts have been done to capture PU before skin breakdown. For instance, thermographic PU studies have demonstrated a temperature rise within the reactive hyperemic phase postloading. They have also found that some patients have a decrease in temperature response, and others, no temperature change upon removal of pressure, suggesting lack of hemodynamic recovery [28]. A model that identifies lack of temperature fluctuations with loading/offloading may be just as significant as one that demonstrates temperature change. Thermographic studies have also demonstrated that vascular recovery post load varies across populations, as well as with the duration and extent of the load. Clinical implementation of thermography has been however limited due to equipment that is cumbersome, expensive, and with relative low accuracy. Furthermore, there is a lack of evidence demonstrating the human response to pressure development over a continuum of time as well as temperature fluctuations in the development of PU in high-risk populations. Developing innovative technologies to continuously monitor markers of PU such as "thermal stress response and recovery" similar to what have been designed for managing DFU [1,28—32] could provide unique insight into stress reperfusion injuries that has not been previously possible in the absence of this technological advance. It will empower the conscious patient to prevent the incident of ulcers via timely alert and notification to shift their weight away from spot at risk. It also empowers administration to better triage patients at risk. It will also empower nursing staff to better identify patient at true risk and implement timely intervention to prevent skin breakdown.

2.3 Venous leg ulcer

The most common type of hard-to-heal wound among the ambulatory older adults is thought to be the venous leg ulcer (VLU) and is also called "venous stasis ulcer" [33]. VLUs are the end stage of venous insufficiency most commonly affects "gaiter region"—the area just above the ankle. Venous insufficiency occurs when the valves within the veins no longer function properly. Valves within veins assist venous blood to go up the leg against gravity. When these valves no longer function properly, blood flows back down the lower extremity and creates a gravity-dependent pressure (e.g., during standing and walking) within the venous system. This leads to elevated pressures within the veins with manifestations of mild-to-severe swelling, aching to frank leg pain, leg heaviness, dilated superficial veins, skin changes, and ultimately ulceration.

VLUs have been estimated to afflict between 0.2% and 1% of the total population and between 1% and 3% of the elderly population in the United States and Europe [1,33−35]. Unfortunately, it is anticipated that with aging of population, the prevalence of VLUs is also increasing in particular among those over 80 years old and in westernized countries [35,36]. Treatment of patients with VLU in the United States is estimated at greater than $2.5 billion per year [37]. Management of VLU includes both nonsurgical and surgical management, wound debridement, and electrical stimulation [1]. It is estimated that 93% of VLUs will heal in 12 months and 7% remain unhealed after 5 years [36]. Similar to DFU, the recurrence of ulcers is high and is estimated to be as high as 70% within 3 months after wound closure [36].

Complications associated with VLU include delayed healing, infections, recurrence, and limited mobility [38] Currently, compression therapy is considered as gold standard [39] for managing VLU. However, compression therapy is underused across numerous geographies [40]. In addition, a significant evidence-practice gap has been reported around the world in appropriate assessment of chronic VLU and timely use of best practice treatments [41]. Some of the key challenges with managing VLUs are lack of clinician knowledge, limited access to specialist multidisciplinary teams, unclear referral pathways, poor communication, local unavailability of compression, and patient unwillingness to receive compression [1,40,41]. Poor patient adherence to compression therapy is another key challenge in effective managing VLUs. The use of technology to improve referral to specialists, providing patient/nurse education, and support to caretaker (e.g., the use of telemedicine), and improve adherence to therapy could address some of the challenges and hurdles for effective management and prevention of VLUs. A promising solution to reduce VLU could be new generation of game-based lower extremity exercise programs (Exergame), which was shown to be effective in improving venous return function and prevent venous stasis. For example, Rahemi et al. [42] demonstrated that 5 min lower extremities Exergame program, which assists in augmenting femoral venous parameters relative to ankle movement and muscle flexion, is effective to improve flow volume, flow velocity, and peak systolic velocity within femoral vein by approximately 50% above baseline. Their study demonstrated that flow volume is remained elevated at least till 15 min post exercise. In their intervention, they designed a simple game, which requires moving a computer cursor as fast as possible from one point to another point by rotating ankle joint. They suggested that this game could be integrated in an mHealth platform for home-based therapy that could assist in preventing deep venous thrombosis and potentially VLUs.

3. Digital health for managing hard-to-heal wounds

The landscape for digital health has evolved through a combination of technology push from instrument and device developers as well as end-user pull from patients, advocacy groups, and more recently providers. In particular, health-related applications embedded in smartphones and wearable monitors (e.g., smartwatches) have become ubiquitous and commoditized. In the following sections, several recent promising digital health developments for preventing hard-to-heal wounds have been summarized. Most of these efforts are however still in infancy and lack clinical validation.

3.1 Digital health to facilitate triaging those at high risk of DFU

Many healthcare quality improvement experts recommend improving the process of high-risk foot care through use of stratified foot risk exams [43]. These exams have been shown to be useful to prevent lower limb amputation (LEA) of DFU up to 70% [44]. However, currently available technologies remain insufficient to be used on a routine basis by nonexpert clinical staff. Unfortunately, the reality is that it has been difficult to translate conventional multifactorial risk stratification models from largely private, tertiary care academic centers to community clinics, which provide care to low-income populations. The centralized center "last resort" referral model is also inherently flawed because central multidisciplinary diabetic foot and wound clinics become easily overloaded when already overtaxed staff and resources become inundated with critically ill patients. This overflow produces scheduling backlogs, increases emergency department visits and hospitalizations, and leads to untimely LEA due to lack of coordinated outpatient risk stratification and prevention of DFU, where intervention should be simpler, effective, and low cost. To fill the gap, new technologies are desperately needed assisting in better risk stratifications without the need of installation of costly infrastructure or training dedicated staff. In an effort to improve ease of classification of plantar wounds, determine risk of LEA, and triage those who could benefit from multidisciplinary care like revascularization, Mills et al. [45] suggested a new wound classification named WIfI, which classifies wound based on three major factors that impact amputation risk and clinical management: wound, ischemia, and foot infection (WIfI). This classification has been adopted by the Society for Vascular Surgery under name of "The Society for Vascular Surgery's Threatened Limb Classification" and integrated in a mobile application, called "SVS iPG," which is freely available for download on Apple App Store or Google Play. The SVS iPG application provides education on different types of wound and includes guidelines to manage each wound type (Fig. 8.3). In addition, it includes a calculator to estimate WIfI score for DFU based on easy-to-assess metrics associated with wound size, ischemia, and degree of infection (Fig. 8.4). Fig. 8.4 illustrates a typical example of WIfI score based on entered information and provides an estimation for risk of LEA as well as a recommendation for benefit of revascularization.

3.2 Digital health to predict and prevent DFU

The best method to manage DFU is preventing it in the first place. Unfortunately, even after healing a DFU, its recurrence rate is very high [2], which may suggest that a DFU is never cured but it is rather in remission. Caring for the patient in remission following a DFU episode has proven challenging under standard practice. Numerous prospective studies have explored how ulcer-free survival is impacted not only by underlying comorbidities, but also by the duration as the patient most recently healed from a DFU episode [17,46]. These investigations suggest that between 30% and 40% of patients experience a recurrent DFU within the first year after healing. Contrary to this, the baseline incidence among all patients with diabetes has been measured between 3.6% and 5.8% [17]. Thus, technologies that could assist in predicting DFU to enable targeted preventative therapies during this critical period may significantly improve patient outcomes and reduce DFU-related costs.

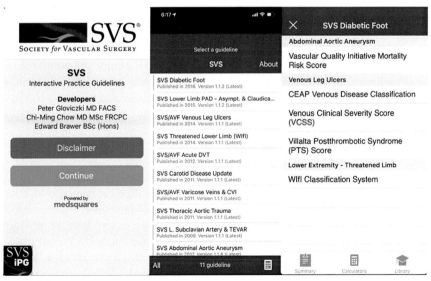

FIGURE 8.3

Society of Vascular Surgery mobile application (SVS iPG app), and interactive practice guideline application for purpose of standardization of care, education, and wound classification. The app is freely available for download on Apple App Store or Google Play.

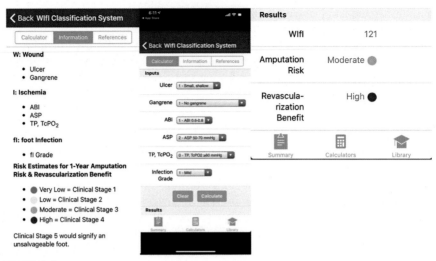

FIGURE 8.4

WIfI calculator application is a part of SVG iPG app (Fig. 8.3) and could be used for classification of lower extremities wounds and consult on standard of care guideline for evaluating the need for multidisciplinary intervention such as potential benefit for revascularization.

Skin temperature measurement and plantar pressure assessments offer objective and reproducible measurement to identify pathologic processes before they result in ulcers [14]. However, a critical issue in implementing these objective assessments for the purpose of routine foot screening application is competing comorbidities for consultant time. Besides clinically based risk classification systems, the addition of sophisticated gait lab measures (e.g., plantar foot pressure and thermometry) have been identified as high-leverage areas for DFU prevention [47–51]. At first glance, the practicality of using these measures under the current constraints of the U.S. healthcare delivery system appears ambitious at best.

Recently, new technologies based on digital health have been developed to address the limitation of previous methods for assessing digital risk factors associated with DFU. In particular, new features such as real-time estimation of risk factors (e.g., mobile-based thermography), using telecommunication means such as smartphone and smartwatch to interact with users and/or caregivers/care providers (e.g., interactive plantar pressure measurement using combination of smart insoles and smartphone or smartwatch), and enhancing form factors to improve level of comfort as well as daily use capability (e.g., using highly flexible sensors, smart textile, etc.) were added to improve ease of measurement of DFU risk factors for the purpose of daily foot monitoring.

3.3 Digital plantar pressure assessment in predicting DFU

In 2017, Raviglione et al. [52] proposed the concept of daily monitoring of plantar pressure in people at risk of DFU using a smart textile (Sensoria socks; Sensoria Inc., Redmond, WA, USA; Fig. 8.5). Their system contained a textile pressure sensor attached to a stretchable band, hardware that collects data and transmits them via Bluetooth to a phone, an app that gathers the data and stores them in the cloud, and a web dashboard that displays the data to the clinician. Their study was however limited to in vitro testing.

Other developed wearable technologies claiming to be able measuring plantar pressure using thin or highly flexible sensors and identified by our systematic search are Walking Sense proposed by Deschamps and Messier in 2015 [53]; smart shoes based on a microstrip patch antenna and able to measure pressure and horizontal force (shear) proposed by Sheibani and colleagues in 2014 [54]; a multiplexed inductive force sensor enables measuring plantar normal and shear forces distribution proposed by Du et al. in 2015 [55]; a highly flexible pressure measuring system based on fiber Bragg grating (FBG) and the lead zirconate titanate piezoceramic sensors, proposed by Suresh et al. in 2015 [56], which enables measuring plantar pressure during walk at low and high speeds; in 2013, Sawacha [57] and Ferber et al. suggested an in-shoe plantar sensory replacement unit to provide alert-based feedback derived from analyzing plantar pressure threshold measurement in real time; in 2008, Dabiri et al. [58] suggested an electronic orthotic shoe that enables monitoring feet motion and pressure distribution beneath the feet in real time and providing continuous feedback (via a text message to the patient and the patient's caregiver) in case of an undesired behavior was detected; a similar concept was proposed in 2008 by Femery et al. [59] in which a biofeedback device (auditory and visual signals) was used to notify excessive, insufficient, or correct foot offloading.

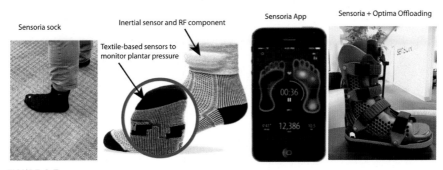

FIGURE 8.5

Sensoria Socks (Sensoria Fitness Inc., Redmond, WA, USA) enables monitoring plantar pressure under three plantar regions of interest including heel, first metatarsal head, and fifth metatarsal head. In addition, it includes an anklet snaps to socks' sensors for transmitting data. Recently Optima and Sensoria teamed up to design a smart offloading boot facilitate real-time assessment of efficacy of offloading.

However, all of these studies were conducted in vitro or using healthy subjects, thus their capability to predict or prevent DFU still needs to be examined.

Some recent studies suggest that real-time notification of harmful plantar pressure is also effective to prevent ulcer recurrence in people with diabetes. Najafi et al. [60] introduced smart insoles to prevent recurrence of ulcers. In their study, 17 patients in diabetic foot remission (history of neuropathic ulceration) were instructed to wear a smart insole system (the SurroSense Rx; Orpyx Medical Technologies Inc., Calgary, Canada; Fig. 8.6) over a 3-month period. This device is designed to cue offloading to manage unprotected sustained plantar pressures in an effort to prevent foot ulceration. The focus of this study was mainly toward the activities that could generate sustained (e.g., greater than 15 min) but not necessarily high pressure such as foot loading during sitting and standing, which could be caused by inappropriate footwear, foreign objects inside of shoes, presence of callus, or prolonged standing. A successful response to an alert was defined as pressure offloading, which occurred within 20 min of the alert onset. Patient adherence, defined as daily hours of device wear, was determined using sensor data and patient questionnaires. Changes in these parameters were assessed monthly. Although no reoccurrence of ulcers was observed in their study, which may suggest that the system was effective to prevent reoccurrence of study, the results however need to be confirmed in a level one evidence with larger sample size. However, an interesting observation in this study is the benefit of real-time alert to improve patient engagement and adherence to prescribed footwear. Specifically, they observed that alerting

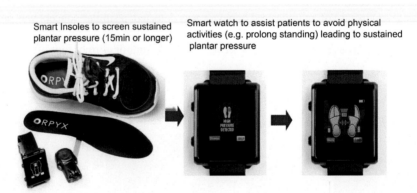

Smart Insoles to screen sustained plantar pressure (15min or longer)

Smart watch to assist patients to avoid physical activities (e.g. prolong standing) leading to sustained plantar pressure

FIGURE 8.6

Recent advances in wearables enable providing timely and real-time feedback to patients to protect their feet against conditions that may increase risk of diabetic foot ulcer. The SurroSense Rx (Orpyx Medical Technologies Inc., Calgary, Canada) is an example of such technologies that enables continuous screening plantar pressure through smart insoles, and notifying the patient in case of a sustained plantar pressure beyond a predefined threshold was detected. This technology could be used to educate patient to avoid conditions leading to sustained plantar pressure (e.g., unbroken and prolonged standing), which in turn could assist with prevention of diabetic foot ulcers.

patients at least once every 2 hours could significantly improve adherence to pre-scribed footwear over time. SurroSense by Orpyx is not the only technology, which provides real-time feedback about plantar pressure.

There are several other mHealth products available in the market that facilitate home monitoring of plantar pressure and gait. These products could be also used for assessing and predicting DFU. FeetMe (FeetMe, Paris, France; Fig. 8.7) is one of those mHealth technologies that instead of using smartwatch as a mean of communication, proposed smartphone to monitor high plantar pressure. However, the clinical validation of these products as well as their acceptability for daily screening of plantar pressure is still unclear.

3.4 Digital plantar temperature assessment in predicting and preventing DFU

In addition to the measurement of plantar pressure, several technologies have been designed to identify preulceration damage to the plantar tissue of the feet via assess-ing plantar temperature [48,50,61,62] or assessing thermal stress response [28,31]. The rational for measuring plantar temperature is built on the notion that the foot heats up before it breaks down into ulcers. The isolated plantar regions with increased heat is mainly due to inflammation response of near damaged or damaged tissue and originally was suggested by Paul Brand and his team in 1975 as an effec-tive method to predict foot ulcer prior skin breakdown [63]. Almost 2 decades later in 1997, Armstrong, Lavery and their colleagues [61] have proposed the use of a portable handheld infrared skin temperature probe as an effective technique to pre-dict foot injury as well as foot complications because of Charcot's arthropathy. How-ever, only recently measuring plantar temperature is becoming feasible as a part of routine clinical assessments to determine foot at risk of complication (e.g., foot ulcer or Charcot) thanks to new generation of digital thermography systems [64]. FLIR

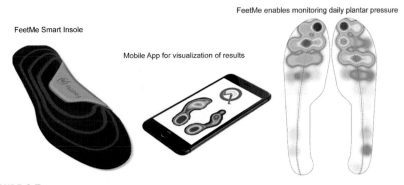

FeetMe Smart Insole

FeetMe enables monitoring daily plantar pressure

Mobile App for visualization of results

FIGURE 8.7

FeetMe devices combine embedded and miniaturized sensors with smart algorithms to assess mobility and plantar pressure.

thermal camera (Fig. 8.8) is among many recent developments for quick and low-cost assessment of plantar temperature, which has been shown to be feasible for assessing diabetic foot at risk of ulceration [64].

Daily assessment of plantar temperature could be also beneficial to prevent DFU. In 2007, Lavery et al. [50] suggested that using thermography as a self-assessment tool is also effective to prevent recurrence of DFU. In their study, physician-blinded, randomized, 15-month, multicenter trial was designed in which 172 subjects with a previous history of DFU were assigned to standard therapy, structured foot examination, or enhanced therapy groups. Each group received therapeutic footwear, diabetic foot education, and regular foot care. Subjects in the structured foot examination group performed a structured foot inspection daily and recorded their findings in a logbook. If standard therapy or structured foot examinations identified any foot abnormalities, subjects were instructed to contact the study nurse immediately. Subjects in the enhanced therapy group used an infrared skin thermometer to measure temperatures on six foot sites each day. Temperature differences >4°F (>2.2°C) between left and right corresponding sites triggered patients to contact the study nurse and reduce activity until temperatures normalized. Their results revealed that patients in the standard therapy and structured foot examination groups were 4.37 and 4.71 times more likely to develop ulcers than patients in the enhanced therapy group. Thus, the study concluded that home monitoring of plantar temperature assessment is a simple and useful tool for prevention of recurrence of ulcers.

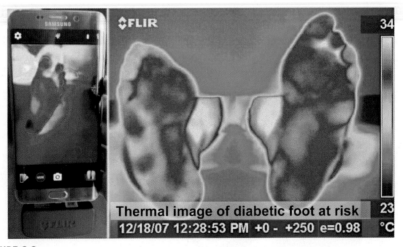

FIGURE 8.8

FLIR thermal camera is installable in a smartphone and enables a quick and low-cost method for determining diabetic foot at risk of ulceration via thermography. By assessing the magnitude of increase in plantar temperature because of stress (e.g., walking) or comparing plantar temperature between feet for two identical plantar regions, could be used as a biomarker of inflammation and predicting skin breakdown.

Despite these studies, simplification of thermography using new generation of mHealth thermography, and clear evidence on benefit of daily plantar temperature assessment, to date, plantar temperature monitoring is still not part of preventive care for managing the diabetic foot. This could be partially because of the form factor of initial thermography tool, which is dependent on patient adherence in daily use of thermography devices as well as the feasibility of an accurate assessment of plantar temperature by nontech savvy patents and their caregivers, and ability to interpret the temperature difference. Almost 2 decades later in 2017, Frykberg et al. [32] proposed a smart mat based on telehealth concept, which could address the limitations of previous thermography tools. Specifically, they studied a novel in−home connected foot mat (Podimetrics Mat, MA, USA; Fig. 8.9) to predict risk of DFU and better stratify those who need an urgent foot care. This simple-to-use system was designed to require no configuration or setup by the users who simply had to step on the mat with both feet for ∼20 s. The system then compared the temperature profile of the two feet. Using a threshold of ≥2.22°C difference between corresponding sites on opposite feet, the mat correctly predicted 97% of DFU with an average lead time of 37 days. Adherence to the mat was high with 86% of participants using the mat at least 3 times per week, and average use was 5 times per week. Although this accuracy and 37 days lead time could be sufficient to better target those who need an urgent care, the technology still suffers from an important limitation. The threshold of 2.22°C, which provided 97% sensitivity, yielded only 43% specificity. Increasing the threshold value resulted in improvements in specificity paired with declines in sensitivity. Under the standard intervention paradigm for monitoring temperatures of the feet, false positives would result in a user being unnecessarily advised to reduce their physical activity level or overwhelm the podiatry clinic with unnecessary patient visits.

Although Podimetrics mat enables daily monitoring of plantar temperature, it provides however a snapshot of plantar temperature irrespective of daily physical activities. Although a single snap shot of temperature could be sufficient to determine sign of inflammation before breakdown of skin or foot ulcers, continuous monitoring

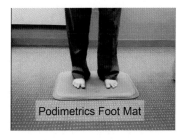

Podimetrics Foot Mat

17°C   20°C

A typical case with asymmetry of plantar temperature captured by the mat

FIGURE 8.9

Podimetrics Mat (Podimetrics Mat, MA, USA) enables the prediction of foot ulceration by facilitating home measurement of plantar temperature. The present algorithm can now identify people with 97% accuracy and an average lead time of 37 days.

of plantar temperature could provide additional information, which may be also useful for reducing false positives, increasing sensitivity, and monitoring other risk factors like shear stress [28,30,31]. In an effort to address this limitation, Najafi et al. [30] designed, fabricated, and evaluated a smart-textile device (SmartSox; Fig. 8.10), which enables simultaneous and continuous measurement of plantar temperature, pressure, and joint angles by measuring changes in light wavelengths circulating in a thin and highly flexible fiber optic weaved in a standard sock. To demonstrate the validity of this device to determine risk factors associated with DFU, they have recruited 33 patients with diabetes and neuropathy attending an outpatient podiatry clinic for regular foot care. All participants were asked to walk a distance of 20 m with their regular shoes at their habitual speed while wearing a pair of SmartSox. To validate the device, results from thermal stress response [28,31] using thermography and peak pressure measured by a computerized pressure insoles (F-scan; TekScan, MA, USA) were used as gold standards. They reported a significant correlation for pressure profile under different anatomical regions of interest between SmartSox and F-scan (r = 0.67, $P < .050$) as well as between thermography and SmartSox (r = 0.55, $P < .050$). In laboratory-controlled conditions,

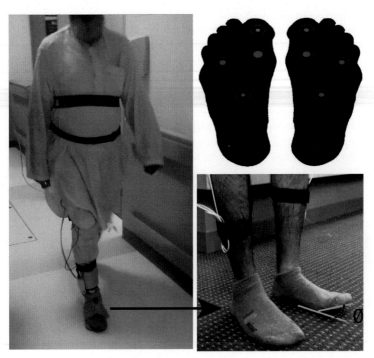

FIGURE 8.10

SmartSox (Najafi et al., 2017) is a smart textile device based on fiber optics enables simultaneous and continuous measurement of plantar pressure, plantar temperature, and lower extremities joint angles (e.g., hallux flexion—extension).

the agreements for parameters of interest were excellent (r > 0.98, *P* < .001). Although, their results are promising and their proposed form factor (using socks instead of insoles) enables monitoring major biomechanical risk factors associated with DFU irrespective of type of footwear (e.g., sandals, offloading, and shoes), it is still unclear whether the proposed measurements are sensitive enough to predict DFU and whether simultaneous and continuous measurements of temperature, pressure, and lower extremity joints angles could add beyond of just snapshot of those measurement like what has been proposed by Podimetrics smart mat (Fig. 8.9) such as reducing false prediction of DFU. Most importantly, the technology proposed by Najafi et al. is not suitable for home assessment as it requires a source of laser light, which required a bulky amplifier that is not small enough to be fully wireless and battery powered.

In 2018, Reyzelman and coworkers proposed a simpler smart socks solution, which could be more appropriate for continuous monitoring of plantar temperature at home and during daily physical activities. The technology named Siren Care Socks (Siren Diabetic Socks; Neurofabric, Siren Care Inc., San Francisco, CA; Fig. 8.11) are made of smart textile with microsensors woven into fabric of socks. These virtually invisible sensors are seamlessly integrated into the socks to monitor temperature changes on the bottom of the feet and under five plantar regions of interest including hallux, metatarsal points 1, 3, and 5, midfoot, as well as heel. The sensor-embedded socks are designed to be reusable and are machine washable and dryable. The device also enables recording lower extremities motion thanks to its triaxial inertial sensor integrated in the shin. Using a mobile application, the plantar temperature could be visualized on daily basis and warning sign is generated based on changes in pattern of plantar temperature. The socks are wirelessly connected with the smartphone via Bluetooth. In a pilot study, authors examined accuracy of the device in 35 people with DPN within three groups: Group 1 included subjects with DPN and no previous history of ulcers (n = 11); Group 2 included

FIGURE 8.11

Siren Socks enables continuous home temperature monitoring, which may be used as an early warning system, to provide people with objective feedback so that they can modify their activity and protect their foot before ulcers develop.

subjects with DPN and a previous history of ulcers (n = 13); and Group 3 included subjects with DPN and a current preulcer as determined by the investigator (n = 11). Participants provide with the socks and were given an Android smartphone with the app needed for temperature monitoring. They were instructed to wear the socks continuously for 6 h, after which the socks could be removed. The recording was continued for 7 ± 2 days. Participants were then requested to grade their experience including usability of the sensor-embedded socks using a questionnaire. Before implementation, the measured temperature by socks was compared to the reference standard, a high-precision thermostatic water bath in the range 20–40°C. Their results suggest that the temperatures measured by the sensors woven into the socks were within 0.2°C of the reference standard with high agreement with the reference system, when the temperature changed between 20°C and 40°C. Patients reported that the socks were easy to use and comfortable, ranking them at a median score of 9 or 10 for comfort and ease of use on a 10-point scale. This study was however limited to feasibility study and validity of the socks to predict foot ulcers needs to be validated in future studies.

There are also number of technologies that enable temperature measurement between insole and shoes, which may also assist in prevention of DFU by indirect measurement of shear stress and generated sweating during daily physical activities. In 2016, Sandoval-Palomares and colleagues [65] suggested a continuous monitoring portable device that monitors the microclimate temperature and humidity of areas between the insole and sole of the footwear. For assessment of their study, two healthy subjects were recruited and were asked to wear a pair of insoles with 1 cm of thickness, which included five sensors integrated in circular holes. The holes created an air channel of microclimate area between the plantar surface-insole and the sensor, which enables reading the plantar surface temperature and humidity data. They have been able to continuously monitor temperature and humidity data without any technical difficulties. However, no gold standard was used to examine the accuracy of reading.

3.5 Digital management of physical activity dosages to prevent or heal DFUs

Identification of specific thresholds for volume of foot stress (e.g., daily number of steps or time spending in standing) that the diabetic foot can tolerate is poorly understood [11,66]. Repetitive pressure and shear stress applied to the surface of the foot as a result of physical activity could lead to inflammation and eventually damage of the soft tissue, which is often unnoticed by patients with insensate foot and consequently could lead to foot ulcer. Similarly, in those with active DFU, excessive foot loading could cause a delay with healing of DFU because of repetitive moderate stress. Therefore, clinicians are often cautious about advising extra activity to their patients with DFU or at risk of DFU [11]. However, several studies have shown that the individuals that ulcerated had lower physical activity levels than those that did not ulcerate [67,68]. Furthermore, while walking on an unprotected wound may be plausibly detrimental to healing, the role of exercise and daily physical activities on health benefits cannot be ignored, even in patients with DFU. On the other hand, prolong immobilization of foot may lead to deconditioning, muscle atrophy and weakness increasing likelihood of frailty and long-term complication for patients [11,69,70].

Armstrong and Boulton [71] suggested that it might be possible to prescribe physical activity "dosages" to at-risk patients. Although standardized physical activity recommendations to reduce DFU risk that are based on cumulative tissue stress remain elusive due to individual differences in stress tolerance, there are opportunities to incorporate technologies into offloading footwear to monitor and/or prescribe personalized modifications to physical activity [5,60,66,70]. Efforts are currently underway to commercialize smart footwear with capability of real-time plantar pressure screening (smart insoles or smart textile) during both static (e.g., standing or sustained pressure) [60] and dynamic (e.g., walking) [52,72] stress applied to the feet and providing this information to both patients and care providers. Some recent efforts have also added a real-time feedback from plantar pressure during activities of daily living that could notify patients about harmful activities. For instance, Raviglione and colleagues [52] introduced a smart textile-based system (Sensoria Inc., Redmond, WA, USA; Fig. 8.5) capable of continuously measuring the pressure and notified high pressure instance via a smartphone. They concluded that this technology could determine optimal offloading in the community setting and assist with prevention of DFU. However, their study was limited to a proof-of-concept design and no clinical study was conducted to support the conclusion.

In addition, some recent research postulated harmful physical activities that could increase the likelihood of foot ulcers. For example, a 2010 study by Najafi et al. [73] found that patients at risk of DFU spent twice as much time per day loading their feet while standing still as opposed to walking. Standing period, often neglected in previous studies focused on managing plantar pressure [11], could be even more risky than walking in people with risk of DFU because of reduced hyperemic response under the diabetic neuropathic foot [14]. Newrick and colleagues [74] have demonstrated that the time for blood flow return to baseline (resting flow) following a 3 min standing is almost twice longer in neuropaths compared to healthy subjects, which could lead to local ischemia and development of callus and consequently an ulcer. Thus, it could be postulated that a prolonged standing period (e.g., longer than 3 min) could be harmful for neuropathic patients and should be avoided. This threshold could be used by new interactive technologies to alert patients when such harmful activities have occurred.

Armstrong et al. [75] has revealed another potential harmful activity behavior, which could increase risk of DFU. The group utilized computerized activity monitors to prospectively monitor the physical activity patterns of 100 patients at risk of DFU. They revealed that those who ulcerated had lower average daily activity levels as well as greater day-to-day variability in their activity level. Based on these studies, Crews et al. [66] concluded in a review paper that low physical activity levels paired with "periodic bursts" in activity could lead to DFU formation. For this conclusion, they inspired from tissue stress theory suggested by Kluding et al.

[76] in which it was suggested that the skin and soft tissue of the feet develop "decreased stress tolerance" in individuals at risk of DFU who maintain low daily activity levels. Thus, when those individuals subsequently have a spike in activity (increased variability) their foot is unable to tolerate the increased stress and a DFU form. In contrast, individuals who maintain consistently higher levels of physical activity have tissue that is conditioned to tolerate a higher level of stress. Under the theory, individuals who gradually increase their physical activity should benefit from gradually increased stress tolerance. However, individuals who excessively exceed their normal physical stress levels will encounter tissue injury or death. Smart shoes, smart insoles, smart socks, or new generation of activity monitoring could assist patients with DFU to maintain a steady activity patterns and avoid an activity burst, which may increase risk of DFU formation.

Management of dose of activities, when patients have active foot ulcers could be also important to avoid delay in wound healing. However, the current literatures are unclear on this topic. Conventional wisdom dictates that once an individual exceeds their tissue's tolerance level and develops a DFU, they should limit activity as much as possible to allow the DFU to heal. However, Vileikyte et al. [77] found that daily step count and offloading adherence level were each positively associated with wound healing in subjects with DFU, while the adherence was stronger predictor of wound healing. This observation is contradictory with observation in Najafi et al., in which 49 patients with active DFU was randomized to receive either a removable cast walker (RCW; with uncertain adherence to offloading during everyday activities) or the same cast walker rendered irremovable (iTCC; with 100% adherence to offloading during day and night time) for the offloading of their wounds. They found negative associations between daily number of taken steps and speed of weekly wound healing irrespective of level of adherence to offloading (RCW or iTCC). They have also reported that the association between low pressure activities (e.g., standing) and speed of wound healing is highly depending on the type of offloading. For instance, those who are using RCW with uncertainly in adherence, a relatively high negative correlation ($r = -0.60$) was observed between duration of standing and weekly speed of wound healing. However, in those who are using iTCC, which reinforced adherence during day and night, no correlation was found between standing and wound healing speed. Interestingly, when they look at the success of wound healing at 12 weeks in a multiple variable model, only standing duration, type of offloading, and initial wound size were independent predictors of success of wound healing, suggesting the importance of managing prolonged and unbroken standing bouts irrespective of type of offloading.

3.6 Digital patient education for managing DFU

Disease education with patients at risk of DFU is a long advocated component of DFU prevention and critical intervention to benefit outcomes [78]. The role of ulcer prevention education is particularly important for prevention of ulcer recurrence immediately after healing of the ulcer, in whom the recurrence rates is estimated to be approximately 40% during the first year [2]. Several reviews and systematic reviews have provided evidence for effectiveness of diabetic education programs in reducing DFU, amputation, and generate cost-effective reduction in lower extremity complications [78—81].

One of the important components of any patient education programs for prevention of DFU is patient adherence to diabetic footwear designed for improving wound healing and/or prevention of diabetic foot ulcers. Clinical care for diabetic ulceration focuses on external offloading of the foot or shifting plantar pressure during gait from the wound area to an unaffected area. However, clinical footwear trials are equivocal, and approximately 40% of these patients still reulcerate within 1 year [2]. A lack of adherence leads to most of these reulcerations. Specifically, despite taking over 50% of their steps at home, patients view their home as "safe zone" where they do not feel the need to wear their prescribed footwear [82]. As a result, high-risk patients wear their prescribed footwear only 15%—28% of the time. Advances in technology allow to implement timely alert, notification, or auto-reminder program to improve adherence.

In 2017, Najafi et al. [60] tested effectiveness of a real-time alerting system (using smartwatch) to improve adherence to prescribed diabetic shoes over time. Participants were asked to wear on daily basis a pair of diabetic shoes equipment with a thin (<0.5 mm) smart insole system (the SurroSense Rx; Orpyx Medical Technologies Inc., Calgary, Canada; Fig. 8.6), over a 3-month period. This device is designed to cue in real-time unprotected sustained plantar pressures. A smartwatch was used to notify in real-time detected sustained pressure. The feedback was provided via watch vibration as well as visual indication of high-pressure plantar spots (with red color) via colorful interface of the smartwatch. Adherence was defined by daily hours of device wear measured objectively by the system. Overall daily adherence was overaged on monthly basis. They reported that those who were received at least one alert every 2 hours had significant improvement in adherence in the month 3 compared to the first month of monitoring. In addition, those who received frequently alerts, improved their success of response to alert, which was defined as successful offloading within 20 min of the alert onset. They concluded that a real-time and comprehensive alert method with a minimum number of alerts (one every 2 hours) are effective to optimally respond to offloading cues from a smart insole system and improving adherence to prescribed diabetic shoes over time.

Some researchers have also proposed the use of smartwatches and smartphones to engage patients in wearing their offloading device. PAMTag (Fig. 8.12) is one of these technologies introduced by Najafi et al. [83] to improve adherence to offloading. The platform includes a smartwatch and a smart tag named PAMTag. PAMTag is attached to an offloading device. The PAMTag includes an accelerometer to monitor activity, an RFID tag, and RF component. The smartwatch is programmed to monitor weight-bearing activities. In the case of detected weight-bearing activities (e.g., standing and walking), the smartwatch communicates with the PAMTag. If PAMTag was not near and/or does not confirmed the offloading was worn during those activities, a notification is provided to the patient via the smartwatch and text

PAMTag

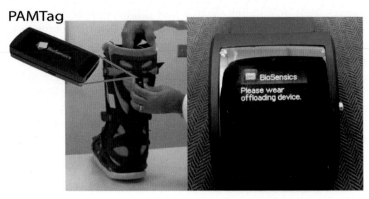

FIGURE 8.12

PAMTag introduced by Najafi et al. and commercialized by Biosensics LLC to engage patients in wearing their offloading devices.

message is sent to patient care givers. In addition, the platform enables reporting daily adherence to offloading to care providers. The validity of this platform to improve adherence to offloading is however unclear.

3.7 Digital health for managing wounds

Smartphones and other consumer digital technologies have emerged as potentially powerful tools to empower patients to take care of their own chronic condition from patient education to engaging them in their own care, monitoring the risk of DFU, and determining any complication associated with wound healing. However, many of these technologies are still in infancy stage. In 2015, Parmanto et al. [84] proposed a smartphone app to support self-skincare tasks, skin condition monitoring, adherence to self-care regimens, skincare consultation, and secure two-way communications between patients and clinicians. The system may help in supporting self-care and adherence to care management, while facilitating communication between patients and clinicians. Wang et al. [85] developed an app for analyzing wound images. The developed app enables capturing wound image with the assistance of an image capture box. The software allows detecting wound boundary and determining healing status. Mammas et al. [86] proposed smartphone as mobile-telemedicine platform. They evaluated the feasibility and reliability of the platform based on simulating experimentation by 10 specialists, who remotely examined a diabetic foot using the proposed mobile platform. They demonstrated that this platform allows remote classification of the wound as well as the risk of amputation with accuracy of 89% on average. In addition, the acceptability of the platform was in range of 89%−100% among specialists. A similar concept was proposed by Foltynski et al. [87] in which an app was designed to measure wound area and sends the data to a clinical database and creates a graph of the wound area change over time. The team has also suggested an elliptical method [88] to improve wound size estimation from 16 different wound shapes. Sanger et al. [89] proposed a smartphone app to engage patients in wound tracking, which in turn could assist in identifying wound infection. However, their study was limited to design concept with no clinical study. An interesting application of mHealth was proposed by Quinn et al. [90] to improve patient referral strategy from tertiary centers. Specifically, they proposed utilizing smartphone technology to decentralize care from tertiary centers to the community, improving efficiency and patient satisfaction, while maintaining patient safety. Their designed app enables remote collection of patient wound images prospectively and their transmission with clinical queries, from the primary healthcare team to the tertiary center. They tested this platform in five public health nurses in geographically remote areas of the region. They demonstrated that images could be transmitted securely, is safe and reliable and could be used for remote wound bed assessment and determine skin integrity/color. They concluded that with minor adjustments, this application could be used across the community to reduce patient attendances at vascular outpatient clinics while still maintaining active tertiary specialist input to their care. Innovative solutions have been also suggested to engage people in the self-care of their diabetic foot ulcers. To facilitate capturing a high-quality picture from plantar wounds, van Netten et al. [91,92] developed and evaluated the usability and usefulness of a smartphone application with voice assistant called "MyFootCare" (Fig. 8.13). Their results suggested that using voice assistance is beneficial to engage patient in tracking their plantar wound size and measuring wound size data is useful to monitor progress and engage people.

3.8 Internet of things and remote management of DFU

One of the fast developing infrastructures promising to revolutionize the wound care industry is the Internet of Things (IoT) [93]. It is expected that up to 50% of healthcare over the next few years will be delivered through virtual platforms. This has accelerated development of a new market named "digital wellness," which combines digital technology and healthcare [93]. Digital technology-based healthcare is

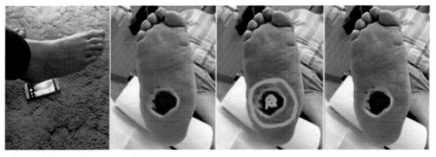

FIGURE 8.13

MyFootCare application developed by van Netten et al. to facilitate capturing high quality photo from plantar wounds using voice-assistance application.

regarded as a natural and ultimate choice for remote, home-based, and long-term care of patients with chronic conditions due to its low cost, high accuracy and continuous monitoring and tracking capabilities. IoT involves a system of devices, machines, or anything with the ability to transfer data without the need for a human to implement the communication [94]. Fueled by the recent adaptation of a variety of enabling wireless technologies such as RFID tags, wearable sensor, and actuator nodes, the IoT has stepped out of its infancy and is the next revolutionary technology in transforming the Internet into a fully integrated "Future Internet" [94]. As we move from www (static pages web) to web2 (social networking web) to web3 (ubiquitous computing web), the need for data-on-demand using sophisticated intuitive queries is increasing significantly. What has made IoT the next big thing is not just its machine-to-machine component, but the potential of sensor-to-machine interaction. With the increasing development of health sensors, there is a growing opportunity to utilize IoT for medical data collection and analysis. It is expected that integration of these tools into the healthcare model has the potential of lowering annual costs of chronic disease management by close to one-third [95]. The use of IoT for medical applications and is however still in infancy. In particular, our systematic search did not identify any study related to application of IoT for wound management. However significant business decisions have been taken recently by major information and communication technology players such as Google, Apple, Cisco, and Amazon to position themselves in the IoT landscape. For example, in 2014 Novartis is working with Google on sensor-technologies, such as the smart lens, and a wearable device to measure blood glucose levels [96]. In 2017, Amazon teamed up with Merck and Luminary Labs on an effort called the Alexa Diabetes Challenge, with the goal of finding the ultimate way to monitor diabetes using voice-enabled solutions [97]. As the IoT continuous to develop, further potential is estimated to develop to facilitate management of chronic conditions at home including effective and timely management of diabetic foot at risk as well as facilitating the delivery of care for speed up wound healing.

4. Conclusion

We live in a world where digital technology is increasingly being integrated into every aspect of our lives. With the miniaturization of processors, advancements in sensing technologies, consistent availability of electrical power, ubiquity of access to the internet, and significant strides in machine learning and artificial intelligence, new emerging solutions have been developed to improve healthcare delivery, patient satisfaction, and population health across different disciplines while reducing the cost of care.

The emergence of "smart" technologies, digital health, and wearable electronics paves the way for the integration of both in the context of providing patients and clinicians with objective data about patient health that is easily accessible. Physicians no longer need to rely on the subjective history given by neuropathic patients who lack the ability to sense the deterioration of their own bodies [98].

Technologies such as mHealth, IoT, and wearables enable cutting down on in-person visits and allow physicians to remotely check in on patients, track patient adherence to the therapy, and detect the early stages of serious medical conditions (e.g., sign of plantar inflammation before breakdown of skin) and triage those who need an immediate supervised or multidisciplinary care. Technology can be used to supplement healthcare provider wound care by providing both educational and motivational support (e.g., gamification and real-time notification systems). The advances in sensing technologies enable the collection of valuable objective and digital markers for hard-to-heal wounds such as pressure/stress, temperature, sustained stress (e.g., unbroken prolonged standing leading to DFUs or long immobility leading to PUs), dosage of weight-bearing activities (e.g., standing and walking), and many more indicators to provide timely intervention to prevent hard-to-heal wounds, while promoting safe mobility. Although the application of such technology for effectiveness of wound care is still in infancy and its cost effectiveness is still debated, by exponential speed of technology development and exponential increase in technology investment for healthcare application, it is anticipated that healthcare and care delivery for chronic conditions such as diabetic foot ulcers, pressure ulcers, and venous leg ulcers will be dramatically changed in a near future.

References

[1] Piaggesi A, Lauchli S, Bassetto F, et al. Advanced therapies in wound management: cell and tissue based therapies, physical and bio-physical therapies smart and IT based technologies. J Wound Care 2018;27(Sup6a):S1−137.

[2] Armstrong DG, Boulton AJM, Bus SA. Diabetic foot ulcers and their recurrence. N Engl J Med 2017;376(24):2367−75.

[3] Lazzarini PA, Pacella RE, Armstrong DG, Van Netten JJ. Diabetes-related lower-extremity complications are a leading cause of the global burden of disability. Diabet Med 2018;35:1297−9.

[4] Lazzarini PA, Hurn SE, Kuys SS, et al. The silent overall burden of foot disease in a representative hospitalised population. Int Wound J 2017;14(4):716−28.

[5] Basatneh R, Najafi B, Armstrong DG. Health sensors, smart home devices, and the internet of medical things: an opportunity for dramatic improvement in care for the lower extremity complications of diabetes. J Diabetes Sci Technol 2018;12(3):577−86.

[6] Barshes NR, Sigireddi M, Wrobel JS, et al. The system of care for the diabetic foot: objectives, outcomes, and opportunities. Diabet Foot Ankle 2013;4.

[7] Skrepnek GH, Mills Sr JL, Armstrong DG. A diabetic emergency one million feet long: disparities and burdens of illness among diabetic foot ulcer cases within emergency departments in the United States, 2006−2010. PLoS One 2015;10(8):e0134914.

[8] Singh N, Armstrong DG, Lipsky BA. Preventing foot ulcers in patients with diabetes. J Am Med Assoc 2005;293(2):217−28.

[9] Rogers LC, Andros G, Caporusso J, Harkless LB, Mills JL, Sr, Armstrong DG. Toe and flow: essential components and structure of the amputation prevention team. J Vasc Surg 2010;52(3 Suppl. 1):23S−7S.

[10] Allen L, Powell-Cope G, Mbah A, Bulat T, Njoh E. A retrospective review of adverse events related to diabetic foot ulcers. Ostomy Wound Manage 2017;63(6):30–3.

[11] Najafi B, Grewal GS, Bharara M, Menzies R, Talal TK, Armstrong DG. Can't stand the pressure: the association between unprotected standing, walking, and wound healing in people with diabetes. J Diabetes Sci Technol 2017;11(4):657–67.

[12] Toosizadeh N, Mohler J, Armstrong DG, Talal TK, Najafi B. The influence of diabetic peripheral neuropathy on local postural muscle and central sensory feedback balance control. PLoS One 2015;10(8):e0135255.

[13] Lavery LA, Hunt NA, Ndip A, Lavery DC, Van Houtum W, Boulton AJ. Impact of chronic kidney disease on survival after amputation in individuals with diabetes. Diabetes Care 2010;33(11):2365–9.

[14] Wrobel JS, Najafi B. Diabetic foot biomechanics and gait dysfunction. J Diabetes Sci Technol 2010;4(4):833–45.

[15] Wrobel JS, Davies ML, Robbins JM. Does open access improve the process and outcome of podiatric care? J Clin Med Res 2011;3(3):101–5.

[16] van Netten JJ, Price PE, Lavery LA, et al. Prevention of foot ulcers in the at-risk patient with diabetes: a systematic review. Diabetes Metabol Res Rev 2016;32:84–98.

[17] Bus SA, van Netten JJ. A shift in priority in diabetic foot care and research: 75% of foot ulcers are preventable. Diabetes Metabol Res Rev 2016;32(Suppl. 1):195–200.

[18] Lyder CH. Pressure ulcer prevention and management. J Am Med Assoc 2003;289(2):223–6.

[19] Schubart JR, Hilgart M, Lyder C. Pressure ulcer prevention and management in spinal cord-injured adults: analysis of educational needs. Adv Skin Wound Care 2008;21(7):322–9.

[20] Schoonhoven L, Grobbee DE, Donders AR, et al. Prediction of pressure ulcer development in hospitalized patients: a tool for risk assessment. Qual Saf Health Care 2006;15(1):65–70.

[21] Pancorbo-Hidalgo PL, Garcia-Fernandez FP, Lopez-Medina IM, Alvarez-Nieto C. Risk assessment scales for pressure ulcer prevention: a systematic review. J Adv Nurs 2006;54(1):94–110.

[22] Whittington KT, Briones R. National Prevalence and Incidence Study: 6-year sequential acute care data. Adv Skin Wound Care 2004;17(9):490–4.

[23] Tubaishat A, Papanikolaou P, Anthony D, Habiballah L. Pressure ulcers prevalence in the acute care setting: a systematic review, 2000–2015. Clin Nurs Res 2018;27(6):643–59.

[24] Gefen A. How much time does it take to get a pressure ulcer? Integrated evidence from human, animal, and in vitro studies. Ostomy Wound Manage 2008;54(10). 26-28, 30-25.

[25] Finestone HM, Levine SP, Carlson GA, Chizinsky KA, Kett RL. Erythema and skin temperature following continuous sitting in spinal cord injured individuals. J Rehabil Res Dev 1991;28(4):27–32.

[26] Edsberg LE. Pressure ulcer tissue histology: an appraisal of current knowledge. Ostomy Wound Manage 2007;53(10):40–9.

[27] Sprigle S, Linden M, McKenna D, Davis K, Riordan B. Clinical skin temperature measurement to predict incipient pressure ulcers. Adv Skin Wound Care 2001;14(3):133–7.

[28] Najafi B, Wrobel JS, Grewal G, et al. Plantar temperature response to walking in diabetes with and without acute Charcot: the Charcot activity response test. J Aging Res 2012;2012:140968.

[29] Wrobel JS, Ammanath P, Le T, et al. A novel shear reduction insole effect on the thermal response to walking stress, balance, and gait. J Diabetes Sci Technol 2014;8(6): 1151—6.

[30] Najafi B, Mohseni H, Grewal GS, Talal TK, Menzies RA, Armstrong DG. An optical-fiber-based smart textile (smart socks) to manage biomechanical risk factors associated with diabetic foot amputation. J Diabetes Sci Technol 2017;11(4):668—77.

[31] Rahemi H, Armstrong DG, Enriquez A, Owl J, Talal TK, Najafi B. Lace up for healthy feet: the impact of shoe closure on plantar stress response. J Diabetes Sci Technol 2017; 11(4):678—84.

[32] Frykberg RG, Gordon IL, Reyzelman AM, et al. Feasibility and efficacy of a smart mat technology to predict development of diabetic plantar ulcers. Diabetes Care 2017;40(7): 973—80.

[33] Margolis DJ, Bilker W, Santanna J, Baumgarten M. Venous leg ulcer: incidence and prevalence in the elderly. J Am Acad Dermatol 2002;46(3):381—6.

[34] Fuhrer MJ, Garber SL, Rintala DH, Clearman R, Hart KA. Pressure ulcers in community-resident persons with spinal cord injury: prevalence and risk factors. Arch Phys Med Rehabil 1993;74(11):1172—7.

[35] Posnett J, Franks PJ. The burden of chronic wounds in the UK. Nurs Times 2008; 104(3):44—5.

[36] Franks PJ, Barker J, Collier M, et al. Management of patients with venous leg ulcers: challenges and current best practice. J Wound Care 2016;25(Suppl. 6):S1—67.

[37] Simka M, Majewski E. The social and economic burden of venous leg ulcers: focus on the role of micronized purified flavonoid fraction adjuvant therapy. Am J Clin Dermatol 2003;4(8):573—81.

[38] Wellborn J, Moceri JT. The lived experiences of persons with chronic venous insufficiency and lower extremity ulcers. J Wound, Ostomy Cont Nurs 2014;41(2):122—6.

[39] O'Donnell Jr TF, Passman MA, Marston WA, et al. Management of venous leg ulcers: clinical practice guidelines of the society for vascular surgery (R) and the American venous forum. J Vasc Surg 2014;60(2 Suppl. l):3S—59S.

[40] Harding K. Challenging passivity in venous leg ulcer care — the ABC model of management. Int Wound J 2016;13(6):1378—84.

[41] Edwards H, Finlayson K, Courtney M, Graves N, Gibb M, Parker C. Health service pathways for patients with chronic leg ulcers: identifying effective pathways for facilitation of evidence based wound care. BMC Health Serv Res 2013;13:86.

[42] Rahemi H, Chung J, Hinko V, et al. Pilot study evaluating the efficacy of exergaming for the prevention of deep venous thrombosis. J Vasc Surg Venous Lymphat Disord 2018; 6(2):146—53.

[43] Hayward RA, Hofer TP, Kerr EA, Krein SL. Quality improvement initiatives: issues in moving from diabetes guidelines to policy. Diabetes Care 2004;27(Suppl. 2):B54—60.

[44] Lavery LA, Wunderlich RP, Tredwell JL. Disease management for the diabetic foot: effectiveness of a diabetic foot prevention program to reduce amputations and hospitalizations. Diabetes Res Clin Pract 2005;70(1):31—7.

[45] Mills Sr JL, Conte MS, Armstrong DG, et al. The society for vascular Surgery lower extremity threatened limb classification system: risk stratification based on wound, ischemia, and foot infection (WIfI). J Vasc Surg 2014;59(1). 220-234 e221-222.

[46] Morbach S, Furchert H, Groblinghoff U, et al. Long-term prognosis of diabetic foot patients and their limbs: amputation and death over the course of a decade. Diabetes Care 2012;35(10):2021—7.

[47] Armstrong DG. Infrared dermal thermometry: the foot and ankle stethoscope. J Foot Ankle Surg 1998;37(1):75−6.

[48] Armstrong DG, Holtz-Neiderer K, Wendel C, Mohler MJ, Kimbriel HR, Lavery LA. Skin temperature monitoring reduces the risk for diabetic foot ulceration in high-risk patients. Am J Med 2007;120(12):1042−6.

[49] Lavery LA, Higgins KR, Lanctot DR, et al. Home monitoring of foot skin temperatures to prevent ulceration. Diabetes Care 2004;27(11):2642−7.

[50] Lavery LA, Higgins KR, Lanctot DR, et al. Preventing diabetic foot ulcer recurrence in high-risk patients: use of temperature monitoring as a self-assessment tool. Diabetes Care 2007;30(1):14−20.

[51] Bus SA, Valk GD, van Deursen RW, et al. Specific guidelines on footwear and offloading. Diabetes Metabol Res Rev 2008;24(Suppl. 1):S192−3.

[52] Raviglione A, Reif R, Macagno M, Vigano D, Schram J, Armstrong D. Real-time smart textile-based system to monitor pressure offloading of diabetic foot ulcers. J Diabetes Sci Technol 2017;11(5):894−8.

[53] Deschamps K, Messier B. Pressure reducing capacity of felt: a feasibility study using a new portable system with thin sensors. Diabetes Res Clin Pract 2015;107(3): e11−14.

[54] Sheibani S, Roshan M, Huang H, Banerjee B, Henderson R. Single chip interrogation system for a smart shoe wireless transponder. In: Paper presented at: engineering in medicine and biology society (EMBC), 2014 36th annual international conference of the IEEE; 2014.

[55] Du L, Zhu X, Zhe J. An inductive sensor for real-time measurement of plantar normal and shear forces distribution. IEEE Trans Biomed Eng 2015;62(5):1316−23.

[56] Suresh R, Bhalla S, Singh C, Kaur N, Hao J, Anand S. Combined application of FBG and PZT sensors for plantar pressure monitoring at low and high speed walking. Technol Health Care 2015;23(1):47−61.

[57] Sawacha Z. Validation of plantar pressure measurements for a novel in-shoe plantar sensory replacement unit. J Diabetes Sci Technol 2013;7(5):1176−8.

[58] Dabiri F, Vahdatpour A, Noshadi H, Hagopian H, Sarrafzadeh M. Electronic orthotics shoe: preventing ulceration in diabetic patients. Conf Proc IEEE Eng Med Biol Soc 2008;2008:771−4.

[59] Femery V, Potdevin F, Thevenon A, Moretto P. Development and test of a new plantar pressure control device for foot unloading. Ann Readapt Med Phys 2008;51(4):231−7.

[60] Najafi B, Ron E, Enriquez A, Marin I, Razjouyan J, Armstrong DG. Smarter sole survival: will neuropathic patients at high risk for ulceration use a smart insole-based foot protection system? J Diabetes Sci Technol 2017;11(4):702−13.

[61] Armstrong DG, Lavery LA, Liswood PJ, Todd WF, Tredwell JA. Infrared dermal thermometry for the high-risk diabetic foot. Phys Ther 1997;77(2):169−75. discussion 176-167.

[62] Frykberg RG, Gordon IL, Reyzelman AM, et al. Feasibility and efficacy of a smart mat technology to predict development of diabetic plantar ulcers. Diabetes Care 2017: dc162294.

[63] Bergtholdt HT, Brand PW. Thermography: an aid in the management of insensitive feet and stumps. Arch Phys Med Rehabil 1975;56(5):205−9.

[64] Fraiwan L, AlKhodari M, Ninan J, Mustafa B, Saleh A, Ghazal M. Diabetic foot ulcer mobile detection system using smart phone thermal camera: a feasibility study. Biomed Eng Online 2017;16(1):117.

[65] Sandoval-Palomares Jde J, Yanez-Mendiola J, Gomez-Espinosa A, Lopez-Vela JM. Portable system for monitoring the microclimate in the footwear-foot interface. Sensors 2016;16(7).

[66] C RT, K AL, Y SV, R. NJ. Recent advances and future opportunities to address challenges in offloading diabetic feet: a mini-review. Gerontology 2018. https://doi.org/10.1159/000486392.

[67] Lemaster JW, Reiber GE, Smith DG, Heagerty PJ, Wallace C. Daily weight-bearing activity does not increase the risk of diabetic foot ulcers. Med Sci Sport Exerc 2003;35(7):1093—9.

[68] Maluf KS, Mueller MJ. Novel Award 2002. Comparison of physical activity and cumulative plantar tissue stress among subjects with and without diabetes mellitus and a history of recurrent plantar ulcers. Clin Biomech 2003;18(7):567—75.

[69] Roser MC, Canavan PK, Najafi B, Cooper Watchman M, Vaishnav K, Armstrong DG. Novel in-shoe exoskeleton for offloading of forefoot pressure for individuals with diabetic foot pathology. J Diabetes Sci Technol 2017;11(5):874—82.

[70] Rahemi H, Nguyen H, Lee H, Najafi B. Toward smart footwear to track frailty phenotypes-using propulsion performance to determine frailty. Sensors 2018;18(6).

[71] DG A, AJ B. Activity monitors: should we begin dosing activity as we dose a drug? J Am Podiatr Med Assoc 2001;9(3):152—3.

[72] Bus SA, Waaijman R, Nollet F. New monitoring technology to objectively assess adherence to prescribed footwear and assistive devices during ambulatory activity. Arch Phys Med Rehabil 2012;93(11):2075—9.

[73] Najafi B, Crews RT, Wrobel JS. Importance of time spent standing for those at risk of diabetic foot ulceration. Diabetes Care 2010;33(11):2448—50.

[74] Newrick PG, Cochrane T, Betts RP, Ward JD, Boulton AJ. Reduced hyperaemic response under the diabetic neuropathic foot. Diabet Med 1988;5(6):570—3.

[75] Armstrong DG, Lavery LA, Holtz-Neiderer K, et al. Variability in activity may precede diabetic foot ulceration. Diabetes Care 2004;27(8):1980—4.

[76] Kluding PM, Bareiss SK, Hastings M, Marcus RL, Sinacore DR, Mueller MJ. Physical training and activity in people with diabetic peripheral neuropathy: paradigm shift. Phys Ther 2017;97(1):31—43.

[77] Vileikyte L, Shen BJ, Brown S, et al. Depression, physical activity, and diabetic foot ulcer healing. American Diabetes Association 77th Scientific Sessions. 2017.

[78] Miller JD, Najafi B, Armstrong DG. Current standards and advances in diabetic ulcer prevention and elderly fall prevention using wearable technology. Curr Geriatrics Rep 2015;4(3):249—56.

[79] O'Meara S, Cullum N, Majid M, Sheldon T. Systematic reviews of wound care management: (3) antimicrobial agents for chronic wounds; (4) diabetic foot ulceration. Health Technol Assess 2000;4(21):1—237.

[80] Dorresteijn JA, Kriegsman DM, Assendelft WJ, Valk GD. Patient education for preventing diabetic foot ulceration. Cochrane Database Syst Rev 2014;12:CD001488.

[81] Crisologo PA, Lavery LA. Remote home monitoring to identify and prevent diabetic foot ulceration. Ann Transl Med 2017;5(21):430.

[82] Armstrong DG, Lavery LA, Kimbriel HR, Nixon BP, Boulton AJ. Activity patterns of patients with diabetic foot ulceration: patients with active ulceration may not adhere to a standard pressure off-loading regimen. Diabetes Care 2003;26(9):2595—7.

[83] Najafi B, Boloori A-R, Wrobel J. Intelligent device to monitor and remind patients with footwear, walking aids, braces, or orthotics. US Patent 8,753,275. 2014.

[84] Parmanto B, Pramana G, Yu DX, Fairman AD, Dicianno BE. Development of mHealth system for supporting self-management and remote consultation of skincare. BMC Med Inf Decis Mak 2015;15:114.

[85] Wang L, Pedersen PC, Strong DM, Tulu B, Agu E, Ignotz R. Smartphone-based wound assessment system for patients with diabetes. IEEE Trans Biomed Eng 2015;62(2): 477—88.

[86] Mammas CS, Geropoulos S, Markou G, Saatsakis G, Lemonidou C, Tentolouris N. Mobile tele-medicine systems in the multidisciplinary approach of diabetes management : the remote prevention of diabetes complications. Stud Health Technol Inf 2014;202: 307—10.

[87] Foltynski P, Ladyzynski P, Wojcicki JM. A new smartphone-based method for wound area measurement. Artif Organs 2014;38(4):346—52.

[88] Foltynski P, Ladyzynski P, Sabalinska S, Wojcicki JM. Accuracy and precision of selected wound area measurement methods in diabetic foot ulceration. Diabetes Technol Ther 2013;15(8):712—21.

[89] Sanger P, Hartzler A, Lober WB, Evans HL, Pratt W. Design considerations for post-acute care mHealth: patient perspectives. AMIA Annu Symp Proc 2014;2014:1920—9.

[90] Quinn EM, Corrigan MA, O'Mullane J, et al. Clinical unity and community empowerment: the use of smartphone technology to empower community management of chronic venous ulcers through the support of a tertiary unit. PLoS One 2013;8(11): e78786.

[91] van Netten JJ, Brown RA, Si da Seng L, Lazzarini PA, Ploderer B. Development of myfootcare: a smartphone application to actively engage people in their diabetic foot ulcer self-Care. 2018.

[92] Ploderer B, Brown R, Seng LSD, Lazzarini PA, van Netten JJ. Promoting self-care of diabetic foot ulcers through a mobile phone app: user-centered design and evaluation. JMIR Diabetes 2018;3(4):e10105.

[93] Murthy DN, Kumar BV. Internet of things (IoT): is IoT a disruptive technology or a disruptive business model? Indian J Mark 2015;45(8):18—27.

[94] Gubbi J, Buyya R, Marusic S, Palaniswami M. Internet of Things (IoT): a vision, architectural elements, and future directions. Future Gener Comput Syst 2013;29(7): 1645—60.

[95] Haughom J. Is the health sensor revolution about to dramatically change healthcare?. 2017.

[96] Senior M. Novartis signs up for Google smart lens. Nat Biotechnol 2014;32(9):856.

[97] Coombs B. How Alexa's best skill could be as a home health-care assistant [Internet]: CNBC; 2017 [cited 1/23/2018]. Podcast. Available from: https://www.cnbc.com/2017/08/09/how-alexas-best-skill-could-be-as-a-home-health-care-assistant.html.

[98] Boghossian JA, Miller JD, Armstrong DG. Towards extending ulcer-free days in remission in the diabetic foot syndrome. In: The diabetic foot syndrome, vol. 26. Karger Publishers; 2018. p. 210—8.

Smartphone-based microscopes

Wenbin Zhu, PhD [1], Cheng Gong[1], Nachiket Kulkarni[1], Christopher David Nguyen[1], Dongkyun Kang, PhD[2]

[1]*College of Optical Sciences, The University of Arizona, Tucson, AZ, United States;* [2]*Assistant Professor, College of Optical Sciences, University of Arizona, Tucson, AZ, United States*

1. Introduction

Disease diagnosis in low-resource settings is often challenging due to the lack of resources and trained personnel. In developed countries, many of the diseases are diagnosed by excision of the suspicious lesions followed by histopathologic assessment. The histopathologic assessment, however, needs dedicated equipment, space, and trained personnel for fixing, sectioning, and staining the tissue. The prepared slides then need to be examined at high magnification under a bench microscope by a pathologist. These resources and trained personnel are often scarce in low-resource settings such as low- and middle-income countries, rural clinics in developed countries, and battlefield clinics. In these settings, disease diagnoses are often made by clinical examination alone, which can lead to inaccurate diagnosis and inadequate treatment [1,2].

Smartphones can potentially play a critical role in improving disease diagnosis in low-resource settings [3,4]. The smartphone works as a "computer in the pocket," where it provides high computing power, easy user interface, and high-definition display. Most relevant to the microscopic imaging, the smartphone has a high-performance camera with a large number of pixels, sufficiently large numerical aperture, and capability to store, analyze, and transfer images. Many of the smartphones are affordable, which facilitated the dissemination and adaptation of the smartphone in low-resource settings. Therefore, one can contemplate developing a microscopy device using a smartphone and examine various biological samples right at the point of care and provide diagnostically relevant information where and when it is most needed.

Recently, there has been exciting development on smartphone-based microscopes with a goal of improving disease diagnosis in low-resource settings. In this chapter, we will review the recent smartphone-based microscopy research. We will first review the general characteristics of the smartphone camera and discuss the advantages of and challenges in using the smartphone camera for microscopy. In the following sections, we will review various imaging modes demonstrated with smartphone microscopes and different types of objective lenses used in smartphone microscopes. Finally, we will discuss the remaining challenges in translating the smartphone microscopy devices into clinically viable diagnostic tools.

2. Smartphone camera

The traditional bench digital microscope includes multilens microscope optics, and an imaging sensor, as well as a computer for image acquisition, display, storage, and further analysis. The smartphone can provide the functionalities provided by the imaging sensor and computer in the bench digital microscope. The smartphone, in conjunction with additional microscopy optics, can be used to construct a standalone digital microscope. To understand the unique benefits of and challenges in using the smartphone camera, we will review specifications of a typical smartphone camera.

The camera lens of the smartphone is specially designed to collect light from a wide angular field and generate images on the imaging sensor with minimal aberrations. Such lenses have F-numbers of 3.0−1.5, which corresponds to numerical apertures (NAs) of 0.17−0.33. NA is defined as the refractive index multiplied with the sine value of the half angle of the focused beam. NA determines the lateral resolution and depth of focus in microscopy. The smartphone camera lens typically has a short focal length, around 3−5 mm, to be incorporated in the thin smartphone. As an example, the iPhone XS smartphone (Apple, CA) uses a camera lens with the F number of 1.8 and focal length of 4.25 mm.

Fig. 9.1 shows the optics design of an example smartphone camera lens [5]. To achieve a large field of view (FOV) with minimal aberrations, multiple elements are used in the smartphone camera lens. Each lens element is typically made with plastic molding process to reduce the cost. The lens aperture is positioned near the front end of the camera. The aperture diameter is typically around a few mm (e.g., estimated aperture diameter of the iPhone XS smartphone: 2.36 mm). As shown in Fig. 9.1, off-axial beams are incident on the imaging sensor with a large incidence angle. The smartphone camera lens is housed in a moving part of the voice-coil motor (VCM). The VCM is used for the focus adjustment of the lens. The focus adjustment capability is useful in the smartphone microscope, which can obviate the needs for an additional focusing mechanism for the microscope objective lens and can allow for easy and interactive focusing adjustment. Inset of Fig. 9.1 shows a photo of an example smartphone camera module (e.g., iPhone SE). The camera module size is less than 10 mm.

The imaging sensor of the smartphone camera has unique features that are beneficial for microscopic imaging. Recent smartphone cameras typically have several megapixels (MP) or more (e.g., iPhone XS imaging sensor: 12 MP), which can provide sufficient sampling resolution for microscopic imaging. The imaging speed of the smartphone imaging sensor is sufficiently high, real-time imaging (e.g., 30 frames/s) or faster, which can be useful for recording dynamic events in the sample. Digital zooming functionality is readily available to further increase the

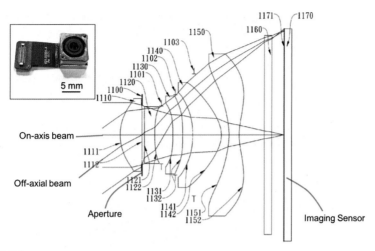

FIGURE 9.1

Schematic of a representative multielement smartphone camera lens. Inset: photograph of a smartphone camera lens module.

Reproduced from US Patent 8.488,259 B2.

sampling resolution and improve the focusing capability on a small region of interest. Some smartphones have optical zooming capability, which can be also used to increase the sampling resolution. Bayer filers are commonly used to generate 24-bit RGB color images (8-bit for each color). Various interpolation methods are used to estimate the missing pixel values in each color channel [6,7]. Use of the color filter reduces the physical number of pixels by two folds for the green channel and by four folds for the red and blue channels (refer to Section 2.2). However, the proprietary interpolation algorithm that each smartphone company uses tends to enhance the edge sharpness and therefore improves the apparent resolution.

Although the large number of pixels of the imaging sensor is beneficial, the imaging sensor typically does not achieve Nyquist sampling in comparison to the optical resolution. With a typical F number of 2, the smartphone camera lens has an NA of 0.25, which produces 1.1 μm resolution, full width half maximum (FWHM) of point spread function (PSF), for a representative wavelength of 520 nm. A typical pixel size of the smartphone camera is around 1.4 μm. Therefore, the pixel size is significantly larger than the Nyquist sampling period, 0.55 μm or half of the resolution, 1.1 μm. The undersampling is even more aggravated when the Bayer color filter is used [8]. When designing the smartphone microscope device, a careful analysis needs to be conducted to achieve adequate sampling.

Standard camera applications usually do not provide functionalities for controlling the exposure time, ISO value, and manual focusing adjustment. There are however free and paid applications (e.g., Open Camera in Android and ProCam in iOS) that allow the user for making these adjustments, which might be necessary for acquiring images at the same condition for quantitative image analysis.

In addition to the high-performance camera, the smartphone provides image display, user interface, data storage, and transfer methods in a compact, low-cost device. This all-in-one portable computer is beneficial for the diagnostic use in low-resource settings, as it can achieve the image storage, analysis, and transfer in one portable device. One potential challenge in the context of the smartphone microscope is that as the microscopy optics and smartphone screen are physically connected, the microscope orientation cannot be adjusted independently from the screen orientation. The microscope arrangement that provides the best image does not always allow for convenient screen orientation for easy image viewing and user control. Additional optics that changes the beam path might be needed to allow for simultaneous, optimal arrangement of the microscope and screen. Smartphones usually have a smaller storage capacity than desktop computers, which might mandate regular downloading of the images to a desktop computer or uploading to a cloud server.

3. Smartphone-based microscopy methods

3.1 Bright field microscopy

Bright field microscopy is the most commonly used microscopy method. Bright field microscopy visualizes changes in light attenuation, scattering, and color of the sample. The sample is either thinly sectioned and placed on the slide glass (in case of the excised tissue sample) or smeared on the slide glass (in case of the blood or fecal sample). In either case, the sample needs to be prepared with the thickness of a few μm. The illumination light is provided from the opposite side of the objective lens (i.e., transillumination). For many biological samples, staining is conducted to mark the structure of interest with a particular color. As an example, hematoxylin and eosin (H&E) staining, most widely used staining method in histopathologic analysis, makes the DNA and RNA appear violet and cytoplasmic proteins red. The illumination light is selectively absorbed by the dye, which lets the remaining wavelengths pass through and generate color in the microscopy images. For unstained samples, the contrast is mainly provided from the light scattering at the edge of membrane structures or other microstructures.

There have been many interesting smartphone bright field microscope devices [9–11]. In the first example, Hutchison et al. have developed a bright field smartphone microscope in conjunction with a microfluidic incubation device to detect *Bacillus anthracis* [10]. When the *B. anthracis* spores ($\sim$ 1 μm size) are present in the sample, they form a distinctive filament structure with the length of tens to hundreds of μm. This transformation from scattered spores into filaments is readily visible with bright field microscopy. Either a 3-mm-diameter glass bead or a 1-mm bead was used as the microscope objective lens to provide the magnification of 100× or 350×. Ambient light was used for the transillumination. Use of the glass bead and no additional light source made the device cost as low as US$0.10 exclusive of the smartphone. The smartphone microscope successfully visualized the filament formation in the case of high-density *B. anthracis* spores while no filaments were observed in the case of high-density spores with a different strain.

In the second example of bright field smartphone microscope, Knowlton et al. have developed a magnetic levitation-based microscope to diagnose sickle cell disease in low-resource settings [11]. Two permanent magnets are used to levitate the blood cells in a capillary (Fig. 9.2A). The blood cells levitate to a height that provides a balance between the magnetic and gravitational forces. The levitation height is dependent on cell density, which can be used to distinguish between healthy and diseased blood cells. An external white LED was used to provide the illumination light. An aspheric singlet with an NA of 0.64 was used as the objective lens. Representative images obtained with this microscope are shown in Fig. 9.2E and F. Sickle cell anemia (SS genotype) blood cells have lower levitation height (Fig. 9.2B) than the control red blood cells (Fig. 9.2C) due to the increased cell density. The image contrast, reduction of the light intensity in the red blood cell, was provided by the light absorption by hemoglobin.

3.2 Dark field microscopy

Dark field microscopy is another commonly used microscopy method. The main difference between bright field and dark field is that in dark field microscopy, the illumination light is configured in such a way that it is not captured by the microscope objective lens when the light is not disturbed by the sample. Only when there is light scattering, some portion of the illumination light is directed toward the microscope objective lens and generates the image contrast. In the bench dark field microscope, the illumination is achieved by using an annular ring, the inner obstruction cone angle of which is larger than the numerical aperture of the microscope objective lens. Dark field microscopy is useful in imaging transparent samples.

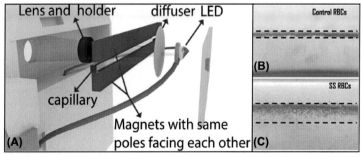

FIGURE 9.2

Brightfield imaging with smartphone microscope. (A) Schematic of a magnetic levitation-based smartphone microscope; (B) Brightfield image of control red blood cells; (C) Brightfield image of sickle cell anemia-infected red blood cells.

Reproduced from Ref. [11], Open Access CCBY 4.0.

Similar to the bench dark field microscope, smartphone dark field microscopes can also use an annular illumination [12,13]. In the smartphone dark field microscope developed by Li, a custom, miniature annular aperture was fabricated using the soft lithography method with liquid metal [12]. The smartphone LED was used as the light source. A half-ball lens with a focal length of 1.81 mm was used as the objective lens. With this microscope, microspheres (diameter = 2.8 μm) were clearly visualized as bright dots (Fig. 9.3B).

A programmable LED array can be used to provide various imaging contrasts [13,14] including dark field contrast. Only the outer ring of the LED array can be activated to provide the annular illumination (green LEDs in Fig. 9.3C). A miniature

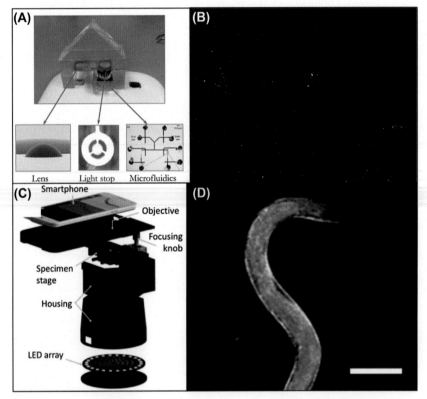

FIGURE 9.3

Dark field imaging with smartphone microscopes. (A) An example smartphone microscope that uses annular illumination mask. (B) Dark field image of microspheres (diameter = 2.8 μm). (C) An example smartphone microscope with programmable LED arrays. (D) Dark field image of a *C. elegans*.

doublet was used as the objective lens. An example image of *Caenorhabditis elegans* (*C. elegans*) is shown in Fig. 9.3D. *C. elegans* is often used as a model organism in various biology studies. The edge and internal space of the *C. elegans* generate bright signals due to light scattering.

Total-internal reflection (TIR)-based dark field smartphone microscopy has been also demonstrated [15]. In this approach, light from a flat-lens LED was coupled into the slide glass. Due to the TIR, the illumination light is mainly confined within the slide glass and can escape the slide glass only when there is a particle on the slide surface. Imaging of gold nanoparticles showed that the TIR-based dark field microscopy method provides better signal to noise ratio than the conventional oblique illumination-based dark field microscopy while obviating the need for the annular illumination aperture.

3.3 Phase imaging

Phase imaging in the bench microscope provides enhanced contrast for transparent samples. The bench microscope conducts the phase imaging by using two different methods: differential interference contrast and phase contrast. Both methods convert the phase difference caused by the refractive index difference into an intensity modulation. To achieve this phase-to-intensity conversion, the bench microscope needs various optical elements to manipulate the optical path length. These optical elements are often expensive and careful alignment of the optical elements is required. Therefore, the traditional phase imaging methods are difficult to implement in smartphone microscopes.

One possible method of achieving phase contrast is by using the aforementioned programmable LED array illumination method [13,14]. The sample can be imaged with a partial illumination from one side of the pupil first and imaged again with a different, partial illumination from the opposite side of the pupil. Then, the difference between the two images can be divided by the sum of the two images to generate a differential phase contrast (DPC). Example DPC images are shown in Fig. 9.4A and (B) Similar to DIC, the DPC images show asymmetric image contrast, one edge of the object exhibits a bright signal while the opposite-side edge dark.

Another method of conducting phase imaging with the smartphone is to acquire multiple images around the focal plane and solve the Poisson equation [16]. Meng et al. have developed a smartphone microscope with a focus adjustment capability. A custom smartphone application was developed to calculate the phase images. For an image with 256×256 pixels, the phase calculation time was ~ 7.4 s. The phase microscope was used to image biological samples such as pap smear.

3.4 Fluorescence imaging

Fluorescence microscopy can visualize specific cellular compounds or molecules by using endogenous or exogenous fluorophores. Fluorescence microscopy in the context of low-resource setting diagnosis can be beneficial in detecting a particular

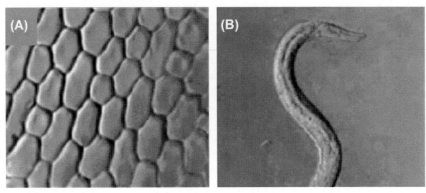

FIGURE 9.4

Phase imaging with smartphone microscope. (A) Differential phase contrast (DPC) image of onion cells; (B) DPC image of a *C. elegans*.

virus, cell, and bacteria in the biopsy specimen or blood sample. There are several requirements needed to realize fluorescence microscopy: (1) excitation light needs to be well matched to the absorption spectrum of the fluorophore of interest; (2) excitation light needs to be mostly blocked in the detection path; and (3) the microscope objective lens NA needs to be high to collect a sufficient number of photons. The first two requirements have already been addressed in Section 2.6.

Using the antibody-conjugated fluorescence dyes, fluorescence microscopy provides molecular specificity. A smartphone epifluorescence microscope was developed to detect specific waterborne parasites in low-resource settings [17]. In this microscope, multiple LEDs were used in conjunction with excitation filters. The LEDs were positioned on the side of the microscopy optics, allowing for imaging of samples mounted on nontransparent sample holders or thick samples. A lens with the focal length of 30 mm was used as the microscope objective lens. A long pass filter was used to reject the back-scattered light from the sample and only allow the fluorescence light detected by the smartphone. The waterborne parasites, *Giardia* cysts, were stained with fluorescein-conjugated antibody. The fluorescence microscopy image clearly visualized *Giardia* cysts over a large FOV, ~ 0.8 cm^2, and a magnified image well distinguished individual cysts.

Fluorescence microscopy can be also used to visualize cell nuclear features from the human patient in vivo or from the freshly excised tissue ex vivo. High-resolution microendoscopy (HRME) is a low-cost fluorescence microscopy technology that can provide such cell nuclear contrast [18,19]. HRME uses a fiber bundle to deliver the excitation light to the tissue and collect the emission light from the tissue while maintaining high spatial resolution. HRME has been demonstrated for imaging various cancerous and precancerous tissues, including cervical cancer, oral cancer, and esophageal cancer [19–21]. As HRME uses a standard fluorescence microscopy optics and electronics, any imaging sensor with a sufficiently low noise level can be used for HRME.

Previously, a DSLR-based HRME was demonstrated [22]. More recently, smartphone-based HRME devices have been demonstrated [23,24]. In one of these devices (Fig. 9.5A), an external LED with a bandpass filter was used to provide the excitation light. Detection light is filtered by another band pass filter. With this device, various tissues and human epithelia were imaged, including the human oral mucosa in vivo (Fig. 9.5B). In this image, the tissue was topically treated with proflavine (0.01% wt/vol in PBS), which preferentially stains the cell nuclei and has a peak excitation wavelength of 445 nm and emission wavelength of 515 nm.

3.5 Reflectance confocal microscopy

Reflectance confocal microscopy (RCM) has been used for imaging various skin diseases in vivo [25]. Different from the fluorescence confocal microscopy commonly used in biology research, RCM uses scattering signals from intrinsic cellular structures of the tissue. As RCM can examine cellular changes associated with the disease status without having to remove the suspicious lesion, it has a potential to enable accurate, timely diagnosis of skin diseases in low-resource settings [26,27]. However, utilization of RCM in low-resource settings has been challenging mainly due to the high device cost and suboptimal portability. Recently, a smartphone-based RCM device was demonstrated [28]. In the smartphone-based confocal microscope (Fig. 9.6A), a diffraction grating is used to investigate multiple line fields simultaneously with each line field linked to a distinctive wavelength. The scan-less approach of conducting confocal imaging made it possible to use the CMOS imaging sensor of the smartphone to capture confocal images. With the smartphone RCM device, characteristic cellular features of the skin were well visualized (Fig. 9.6B and C).

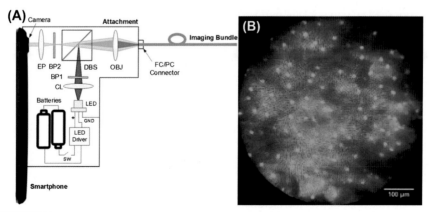

FIGURE 9.5

Fluorescence imaging with smartphone microscope. (A) Smartphone-based high-resolution microendoscope; (B) Fluorescence image of human oral mucosa in vivo.

Reproduced from Ref. [23], Open Access CC.

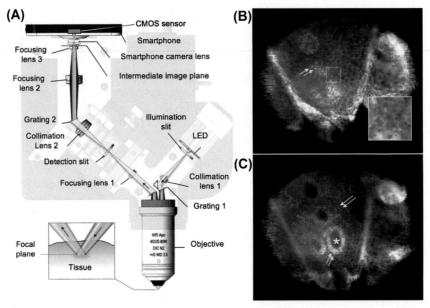

FIGURE 9.6

Reflectance confocal imaging with a smartphone microscope. (A) Schematic of the smartphone confocal microscope; (B) and (C) Confocal images of human skin in vivo. Arrows: melanocytes or melanin-containing basal cells; Asterisk: dermal papillae.

Reproduced from Ref. [28] with permission from OSA.

4. Objective lenses for smartphone-based microscopes

One of the key components in microscopy optics is the objective lens. Although a low-cost lens, such as a ball lens, can provide high magnification and high numerical aperture, other performance specifications might be degraded, such as the FOV and chromatic aberration correction. In this section, we will review three different approaches for the objective lens in the smartphone microscope and discuss the advantages and challenges of each approach.

4.1 Singlet lens (ball lens, half-ball lens, aspheric lens, or liquid lens)

Many smartphone microscopes have used a singlet lens as the objective lens (Fig. 9.7A) [10,12,29,30]. The lens cost is very low, often <US$1, which makes this approach attractive for the smartphone microscope. Ball and half-ball lenses are the most readily available at the lowest cost. The ball and half-ball lenses can have a sufficiently small focal length, which can provide a high numerical aperture needed for microscopic imaging. However, cautions should be taken when estimating the effective NA as the aperture diameter of the smartphone camera lens

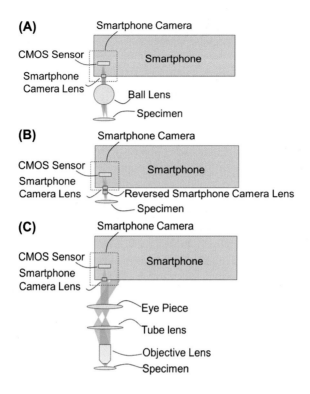

(A)

Smartphone Camera

CMOS Sensor

Smartphone

Smartphone
Camera Lens

Ball Lens

Specimen

(B)

Smartphone Camera

CMOS Sensor

Smartphone

Smartphone
Camera Lens

Reversed Smartphone Camera Lens

Specimen

(C)

Smartphone Camera

CMOS Sensor

Smartphone

Smartphone
Camera Lens

Eye Piece

Tube lens

Objective Lens

Specimen

FIGURE 9.7

Various objective lens approaches in smartphone microscopes: (A) singlet lens;
(B) multielement smartphone camera lens; (C) commercial multielement microscope
objective lens.

is generally smaller than the objective lens diameter. The biggest challenge with the
ball and half-ball lenses is aberrations that these lenses have. Even at the center of
the FOV, the objective lens has significant spherical aberrations. As the field height
increases, the aberrations increase rapidly. Field curvature is another big issue as
most samples imaged by the smartphone microscope are flat. These aberrations
reduce the effective FOV and make the effective resolution worse than the theoretical value expected by the given NA.

An aspheric singlet can provide better spherical aberration reduction [11].
Aspheric singlets are generally used for collimating or coupling the light from the
laser diode and provide sufficiently large NA, over 0.25. Most aspheric lenses are
fabricated by the molding process, making the lens price relatively low for the given
NA, ~US$50. Similar to the ball and half-ball lenses, the small aperture diameter of
the smartphone camera lens should be taken account when calculating the effective
NA. Although the on-axis performance can be enhanced with the aspheric singlet,
the off-axial aberrations are still significant as the aspheric lenses are mainly
designed for the on-axis light collimation or coupling. When multiple colors are

imaged, there are severe chromatic aberrations, different wavelengths focused on different locations, in any singlet approaches.

Lastly, a liquid lens can be used as the objective lens. An objective lens can be fabricated by an inkjet printing method [30,31]. By changing the volume and temperature of the curable liquid, for example, polydimethylsiloxane, the focal length of the lens can be controlled. This provides a flexible platform to fabricate an objective lens with the required focal length at low cost. Another liquid lens approach is to fill an elastomer chamber with a liquid and change the focus by changing the pressure on the liquid [29]. The variable focusing approach can provide a multimagnification imaging platform in the smartphone microscope.

4.2 Multielement smartphone camera lens

As discussed in Section 2, the smartphone camera lens is specially designed to generate images on the relatively large (a few mm) imaging sensor of the smartphone with minimal aberrations. The smartphone camera lens has a moderately high NA, ∼0.25, which provides around 1 μm resolution. Therefore, the smartphone camera lens can be used as an objective lens in certain applications. In this approach [8,32−34], the aperture of the additional smartphone camera lens is placed near the aperture of the camera lens of the smartphone (Fig. 9.7B). The use of two identical smartphone camera lenses constitutes 1:1 imaging optics between the sample and imaging sensor. The large FOV of the smartphone camera lens, a few mm, now becomes the FOV of the smartphone microscope, which is larger than the FOV of any other objective lens approaches. Two potential challenges with this approach, however, are (1) the NA provided is smaller than that of the bench microscope, typically around 0.65, and therefore the resolution and light collection efficiency are limited; and (2) the illumination optics needs to be carefully designed to accommodate the nontelecentric arrangement of the smartphone camera lens aperture.

4.3 Multielement commercial microscope objective lens

Multielement objective lenses used in bench microscopes are well corrected for common aberrations. The smartphone can be integrated with the existing bench microscope to capture images [35−37]. The bench microscope usually uses an objective lens, a tube lens, and an eyepiece lens as shown in Fig. 9.7C. The eyepiece lens is optimized for human eyes with a pupil diameter of 4 mm. The pupil diameter of the smartphone camera lens is typically around 2 mm, and therefore the effective NA of the objective lens is reduced when the smartphone camera is directly mounted on the eyepiece. Another challenge is that the smartphone camera lens aperture is not always positioned at the design pupil plane for the human eye, which generates vignetting of some of the off-axial beams and subsequently reduces the effective FOV. Due to these challenges, to utilize the full NA and FOV of the commercial microscope objective lens, a relay optics needs to be used to map the objective lens pupil diameter to the smartphone camera lens aperture.

4.4 FOV versus resolution

The goal of the smartphone microscope design is to achieve a comparable image performance as the bench microscope or better. We have compared the imaging performances of the reported smartphone microscopes with commercial microscope objective lens performances (Fig. 9.8). When the reference paper describes the measured or expected resolution, that value was used in the plot. When the measured or expected resolution information is missing in the publication, the resolution is calculated as the FWHM of the PSF at 520 nm using the NA provided in the reference. For the commercial objective lenses, data for Olympus UPLAPO lenses with various magnifications and NA were used. FOV was calculated by dividing the field number by the magnification. X-axis shows the resolution in μm and y-axis shows the FOV in mm. Both resolution and FOV are plotted in log scale.

As shown in Fig. 9.8, the reported smartphone microscopy resolutions are generally poorer than commercial microscope objective resolution. Most smartphone microscopes have a lateral resolution larger than 1 μm. The number of resolvable points, FOV divided by resolution, is several-fold lower in the smartphone microscopes, 194—1299 in the smartphone microscopes versus 1499—3997 in commercial microscope objective lenses. It is noted here that many of the smartphone microscope publications miss information on the FOV or resolution or both and therefore are not included in Fig. 9.8. It is also noted that other low-cost microscopy methods of simultaneously achieving large FOV and high resolution such as lens-less microscope and ptychographic microscope are not included here as these methods require

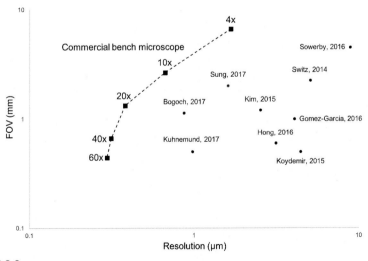

FIGURE 9.8

Comparison of optical performance for various smartphone microscopes and commercial bench microscopes.

significant alternation of the smartphone device itself (e.g., removing the lens from the imaging sensor) or have not been demonstrated in a smartphone device yet.

5. Conclusion

In this chapter, we have reviewed recent works on the smartphone microscope developed for various applications. Thanks to the low cost and ever-improving performance of the smartphone, we expect to see more of the exciting studies on new and improved smartphone microscopes and their utilization in the low-resource settings.

There are, however, several remaining challenges before the smartphone microscopes are widely adapted in clinical settings. First and most critical hurdle to jump over is clinical validation. Most of the published works are focused on developing the smartphone microscopy technology but often lack rigorous evaluation of the diagnostic accuracy in comparison with the standard methods. This validation study will need to include the use of the smartphone microscope in low-resource settings by low-cadre health workers with minimal engineering or medical training.

Achieving high image quality consistently is very important. The smartphone microscope application needs to implement periodic testing of the device performance with the reference sample and means to calibrate the microscope. Real-time image quality checking will need to be included as part of the smartphone application to ensure that all the captured images have acceptable image quality.

One of the major challenges in adapting new microscopy technologies for clinical applications is the requirement for the trained personnel who can interpret the images. Traditionally, microscopy images were read by pathologists or other clinicians with medical training. This approach is not compatible with low-resource settings. Recently, there are several exciting outcomes from studies utilizing artificial intelligence for automatically analyzing microscopy images. Deep learning (DL)-based image analysis method for esophageal optical coherence tomography images has been successfully demonstrated [38] and recently cleared by the US FDA for clinical implementation. DL-based automated image analysis algorithm also has been used for skin microscopy images [39–41]. Although initial development of the DL network is computationally intensive, once the network is developed, it does not require high computing power and can be potentially implemented as part of the smartphone application. The DL approach can be also used to improve image quality even when the original images are obtained with a low-NA objective lens [42].

Finally, most of the smartphone microscopes reported only work with a particular smartphone. For a different smartphone, some degree of design modification is mandated. Most smartphones have similar camera specifications (focal length, F-number, and pixel size in MP) but have different camera location relative to the smartphone body. Therefore, an agile design that needs a minimal change of the mechanical interface between the microscope attachment and smartphone is desirable.

Similarly to the smartphone microscope hardware, the smartphone microscope applications currently need to be developed for a particular platform and is not compatible with other platforms. A simple smartphone application that conducts the image acquisition but works with a cloud server to accomplish the image analysis could be a potential solution.

References

[1] Laker-Oketta MO, Wenger M, Semeere A, Castelnuovo B, Kambugu A, Lukande R, Asirwa FC, Busakhala N, Buziba N, Diero L, et al. Task shifting and skin punch for the histologic diagnosis of Kaposi's sarcoma in sub-saharan Africa: a public health solution to a public health problem. Oncol Times 2015. https://doi.org/10.1159/000375165.

[2] Amerson E, Buziba N, Wabinga H, Wenger M, Bwana M, Muyindike W, Kyakwera C, Laker M, Mbidde E, Yiannoutsos C, et al. Diagnosing Kaposi's sarcoma (KS) in East Africa: how accurate are clinicians and pathologists?. In: 13th Int. Conf. Malig. AIDS other Acquir. Immunodefic. ICMAOI 2011 Bethesda, MD United States; 2012. https://doi.org/10.1186/1750-9378-7-S1-P6.

[3] Kahn JG, Yang JS, Kahn JS. "Mobile" health needs and opportunities in developing countries. Health Aff 2010. https://doi.org/10.1377/hlthaff2009.0965.

[4] Bastawrous A, Armstrong MJ. Mobile health use in low-and high-income countries: an overview of the peer-reviewed literature. J R Soc Med 2013. https://doi.org/10.1177/0141076812472620.

[5] Chen C, Tang H, Hsu P, Hsieh D. Optical image capturing lens system. US 8.488,259 B2. July 2013.

[6] Li X, Orchard MT. New edge-directed interpolation. IEEE Trans Image Process 2001. https://doi.org/10.1109/83.951537.

[7] Lin T. Color interpolator and horizontal/vertical edge enhancer using two line buffer and alternating even/odd filters for digital camera30; 2003. Google Patents.

[8] Switz NA, D'Ambrosio MV, Fletcher DA. Low-cost mobile phone microscopy with a reversed mobile phone camera lens. PLoS One 2014. https://doi.org/10.1371/journal.pone.0095330.

[9] Sowerby SJ, Crump JA, Johnstone MC, Krause KL, Hill PC. Smartphone microscopy of parasite eggs accumulated into a single field of view. Am J Trop Med Hyg 2016. https://doi.org/10.4269/ajtmh.15-0427.

[10] Hutchison JR, Erikson RL, Sheen AM, Ozanich RM, Kelly RT. Reagent-free and portable detection of *Bacillus anthracis* spores using a microfluidic incubator and smartphone microscope. Analyst 2015;140(18):6269−76. https://doi.org/10.1039/c5an01304f.

[11] Knowlton SM, Sencan I, Aytar Y, Khoory J, Heeney MM, Ghiran IC, Tasoglu S. Sickle cell detection using a smartphone. Sci Rep 2015;5(1):15022. https://doi.org/10.1038/srep15022.

[12] Li Z. Miniature optofluidic darkfield microscope for biosensing. In: Liu Z, editor. Ultrafast nonlinear imaging and spectroscopy II, vol. 9198. International Society for Optics and Photonics; 2014. 91980G. https://doi.org/10.1117/12.2062642.

[13] Jung D, Choi JH, Kim S, Ryu S, Lee W, Lee JS, Joo C. Smartphone-based multi-contrast microscope using color-multiplexed illumination. Sci Rep 2017. https://doi.org/10.1038/s41598-017-07703-w.

[14] Phillips ZF, D'Ambrosio MV, Tian L, Rulison JJ, Patel HS, Sadras N, Gande AV, Switz NA, Fletcher DA, Waller L. Multi-contrast imaging and digital refocusing on a mobile microscope with a domed LED array. PLoS One 2015. https://doi.org/10.1371/journal.pone.0124938.

[15] Kim BN, Diaz JA, Gweon Hong S, Hun Lee S, Lee LP. Dark-field smartphone microscope with nanoscale resolution for molecular diagnostics. In: 18th International conference on miniaturized systems for chemistry and life sciences; San Antonio; 2014. p. 2247.

[16] Meng X, Huang H, Yan K, Tian X, Yu W, Cui H, Kong Y, Xue L, Liu C, Wang S. Smartphone based hand-held quantitative phase microscope using the transport of intensity equation method. Lab Chip 2017;17(1):104−9. https://doi.org/10.1039/C6LC01321J.

[17] Koydemir HC, Gorocs Z, Tseng D, Cortazar B, Feng S, Chan RYL, Burbano J, McLeod E, Ozcan A. Rapid imaging, detection and quantification of Giardia lamblia cysts using mobile-phone based fluorescent microscopy and machine learning. Lab Chip 2015. https://doi.org/10.1039/c4lc01358a.

[18] Pierce M, Yu D, Richards-Kortum R. High-resolution fiber-optic microendoscopy for in situ cellular imaging. J Vis Exp 2011;47.

[19] Muldoon TJ, Roblyer D, Williams MD, Stepanek VMT, Richards-Kortum R, Gillenwater AM. Noninvasive imaging of oral neoplasia with a high-resolution fiber-optic microendoscope. Head Neck 2012. https://doi.org/10.1002/hed.21735.

[20] Quinn MK, Bubi TC, Pierce MC, Kayembe MK, Ramogola-Masire D, Richards-Kortum R. High-resolution microendoscopy for the detection of cervical neoplasia in low-resource settings. PLoS One 2012;7(9):e44924.

[21] Muldoon TJ, Thekkek N, Roblyer D, Maru D, Harpaz N, Potack J, Anandasabapathy S, Richards-Kortum R. Evaluation of quantitative image analysis criteria for the high-resolution microendoscopic detection of neoplasia in barrett's Esophagus. J Biomed Opt 2014. https://doi.org/10.1117/1.3406386.

[22] Shin D, Pierce MC, Gillenwater AM, Williams MD, Richards-Kortum RR. A fiber-optic fluorescence microscope using a consumer-grade digital camera for in vivo cellular imaging. PLoS One 2010. https://doi.org/10.1371/journal.pone.0011218.

[23] Hong X, Nagarajan VK, Mugler DH, Yu B. Smartphone microendoscopy for high resolution fluorescence imaging. J Innov Opt Health Sci 2016. https://doi.org/10.1142/s1793545816500462.

[24] Gómez-García PA, Brognara G, Pratavieira S, Kurachi C. Dual configuration clinical fluorescence microscope for smartphones. 2016. https://doi.org/10.1364/cancer.2016.jtu3a.41.

[25] Rajadhyaksha M, Marghoob A, Rossi A, Halpern AC, Nehal KS. Reflectance confocal microscopy of skin in vivo: from bench to bedside. Lasers Surg Med 2017;49(1):7−19.

[26] Corsetti Grazziotin T, Cota C, Bortoli Buffon R, Araújo Pinto L, Latini A, Ardigò M. Preliminary evaluation of in vivo reflectance confocal microscopy features of Kaposi's sarcoma. Dermatology 2010. https://doi.org/10.1159/000297561.

[27] Segura S, Puig S, Carrera C, Lecha M, Borges V, Malvehy J. Non-invasive management of non-melanoma skin cancer in patients with cancer predisposition genodermatosis: a role for confocal microscopy and photodynamic therapy. J Eur Acad Dermatol Venereol 2011;25(7):819−27.

[28] Freeman EE, Semeere A, Osman H, Peterson G, Rajadhyaksha M, González S, Martin JN, Anderson RR, Tearney GJ, Kang D. Smartphone confocal microscopy for imaging cellular structures in human skin in vivo. Biomed Opt Express 2018;9(4): 1906−15. https://doi.org/10.1364/BOE.9.001906.

[29] Huang H, Wei K, Zhao Y. Variable focus smartphone based microscope using an elastomer liquid lens. In: 2016 IEEE 29th International Conference on micro electro mechanical systems (MEMS). IEEE; 2016. p. 808−11. https://doi.org/10.1109/MEMSYS.2016.7421752.

[30] Sung Y-L, Jeang J, Lee C-H, Shih W-C. Fabricating optical lenses by inkjet printing and heat-assisted *in situ* curing of polydimethylsiloxane for smartphone microscopy. J Biomed Opt 2015;20(4):047005. https://doi.org/10.1117/1.JBO.20.4.047005.

[31] Sung Y, Campa F, Shih W-C. Open-source do-it-yourself multi-color fluorescence smartphone microscopy. Biomed Opt Express 2017. https://doi.org/10.1364/boe.8.005075.

[32] Kim JH, Joo HG, Kim TH, Ju YG. A smartphone-based fluorescence microscope utilizing an external phone camera lens module. Biochip J 2015. https://doi.org/10.1007/s13206-015-9403-0.

[33] Kühnemund M, Wei Q, Darai E, Wang Y, Hernández-Neuta I, Yang Z, Tseng D, Ahlford A, Mathot L, Sjöblom T. Targeted DNA sequencing and in situ mutation analysis using mobile phone microscopy. Nat Commun 2017;8:13913.

[34] Bogoch II, Koydemir HC, Tseng D, Ephraim RKD, Duah E, Tee J, Andrews JR, Ozcan A. Evaluation of a mobile phone-based microscope for screening of schistosoma haematobium infection in rural Ghana. Am J Trop Med Hyg 2017. https://doi.org/10.4269/ajtmh.16-0912.

[35] Baek D, Cho S, Yun K, Youn K, Bang H. Time-lapse microscopy using smartphone with Augmented reality markers. Microsc Res Tech 2014;77(4):243−9. https://doi.org/10.1002/jemt.22335.

[36] Roy S, Pantanowitz L, Amin M, Seethala RR, Ishtiaque A, Yousem SA, Parwani AV, Cucoranu I, Hartman DJ. Smartphone adapters for digital photomicrography. J Pathol Inform 2014;5(1):24. https://doi.org/10.4103/2153-3539.137728.

[37] Fontelo P, Liu F, Yagi Y. Evaluation of a smartphone for telepathology: lessons learned. J Pathol Inform 2015;6:35. https://doi.org/10.4103/2153-3539.158912.

[38] Swager AF, van der Sommen F, Klomp SR, Zinger S, Meijer SL, Schoon EJ, Bergman JJGHM, de With PH, Curvers WL. Computer-aided detection of Early barrett's neoplasia using volumetric laser Endomicroscopy. Gastrointest Endosc 2017. https://doi.org/10.1016/j.gie.2017.03.011.

[39] Kurugol S, Kose K, Park B, Dy JG, Brooks DH, Rajadhyaksha M. Automated delineation of dermal−Epidermal junction in reflectance confocal microscopy image stacks of human skin. J Invest Dermatol 2015;135(3):710−7.

[40] Kose K, Alessi-Fox C, Rajadhyaksha MM, Gill M, Brooks DH, Bozkurt A, Dy JG. A multiresolution deep learning framework for automated annotation of reflectance confocal microscopy images. 2018. https://doi.org/10.1364/microscopy.2018.mth2a.1.

[41] Esteva A, Kuprel B, Novoa RA, Ko J, Swetter SM, Blau HM, Thrun S. Dermatologist-level classification of skin cancer with deep neural networks. Nature 2017. https://doi.org/10.1038/nature21056.

[42] Rivenson Y, Göröcs Z, Günaydin H, Zhang Y, Wang H, Ozcan A. Deep learning microscopy. Optica 2017.

Smartphone for monitoring basic vital signs: miniaturized, near-field communication based devices for chronic recording of health

Alex Burton, BS [1], Tucker Stuart, BS [1], Jokubas Ausra, BS [1], Philipp Gutruf, BS, PhD [2]

[1]*Department of Biomedical Engineering, University of Arizona, Tucson, AZ, United States;*
[2]*Assistant Professor, Department of Biomedical Engineering, BIO5 Institute, Department of Electrical Engineering, University of Arizona, Tucson, AZ, United States*

1. Introduction

Within the ever-expanding landscape of the internet of things, medical devices that provide means for chronic recording of biodata, transmission, and centralized analysis can expand diagnostic capabilities well beyond the limitations of current clinical tools and methods. The availability of this data coupled with modern statistical analysis tools such as machine learning can drive a new standard of health diagnostics commonly referred to as digital medicine [1]. To proliferate such technology, there is a need for new sensing hardware that enables wireless, clinical grade, and chronic recording with form factors and device lifetimes that match application length and are minimally invasive to subjects' activities, thus presenting virtually no barrier of usage.

These requirements stand at odds with current device technologies, such as fitness wearables, that either lack high fidelity data streams that can be used for health diagnostics or have form factors and/or operation times that are not suitable for chronic recording. Bridging this gap in device technology, especially in the realm of bio interfaces and sensors, is critical for successful realization of the digital medicine concept [2].

An example of emerging technology aimed at bridging this gap are so-called epidermal electronic devices, a class of soft electronic circuits that are laminated directly onto the skin to obtain an intimate interface with the target organ. These ultrasoft devices that have comparable mechanics to children's temporary tattoos rely on material systems that are virtually identical to high performance circuits and semiconductors, and are backed by a stretchable backbone of plastics and elastomeric materials. These characteristics enable high performance computing and digitalization coupled with biocompatible mechanics, resulting in a high-fidelity bio interface. Examples of such devices that perform superior in comparison to current rigid counterparts are thin epidermal electronics capable of recording electrophysiology [3], macrovascular blood flow [4], and mechanical properties of soft tissue [5]. These demonstrations show the potential of such devices, however, lack sophisticated interfaces that would allow for a facile connection to advanced data collection and analysis tools.

For epidermal devices to be translated into clinically relevant applications, unification of wireless technologies with such platforms is needed. Achieving this task, however, is difficult due to the lack of soft power supplies that match energy needs of contemporary wireless electronic radios, digital electronics, and analog circuits needed for acquisition of biomedical signals. In this chapter, we will introduce possible solutions to this problem by showcasing devices, as schematically shown in Fig. 10.1, that utilize common wireless technology standards, prevalent in most modern smartphones, to power and read epidermal devices. Synthesis of these technologies aims to facilitate a gateway for chronically recorded, high fidelity data streams that can be relayed into the cloud for analysis and diagnostics. Achieving this task will make strides toward realization of the next generation digital medicine devices.

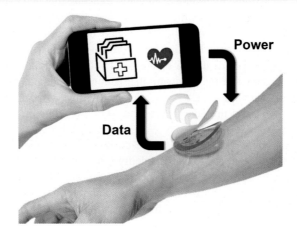

FIGURE 10.1

Concept schematic illustrating the seamless interface of smartphone and epidermal electronics.

2. Introduction to digital and soft battery-free devices with wireless operation

Near-field communication (NFC) technology is implemented in many modern electronics including smartphones, mobile payment, wearables, and access control. This allows new wireless device platforms that incorporate this standard to integrate into a rich infrastructure. Popularity of this technology stems from the ability to operate devices in a fully passive mode using only the reader to supply power. Features include data storage, encryption, and system on a chip (SOC) embodiments featuring NFC frontends that use microcontrollers to introduce a rich set of peripherals such as controllable timers and analog-to-digital converters (ADC).

NFC has the capability to transfer both data and power through electromagnetic coupling between a primary coil (reader) and a secondary coil (NFC-enabled device). When the reader gets in range of the secondary coil, power is sent to the NFC chip, allowing it to initialize a protocol that starts the communication sequence between the two devices. These NFC-enabled devices can be applied to fit many forms of embodiments and perform various sensing tasks, which we will explore in this chapter. The frequency for both power transfer and data communication is fixed to 13.56 MHz. This allows for all NFC compatible devices to communicate without the need for additional licensing when using the system within the specifications of the industrial, scientific, and medical radio band [6,7]. The efficiency of power transfer between the primary and secondary coil is determined by the quality factor (Q factor) of the transmission and secondary coil. Both the spatial alignment and tuning of the coils can be optimized for communication up to 1 m [8].

The high integration density of SOC with NFC technology enables highly miniaturized devices while maintaining a rich feature set. A summary of such capabilities is schematically shown in Fig. 10.2A. Here, the power harvested by the reader powers a microcontroller that can drive peripheral devices such as light sources, heaters, analog amplification circuits, and digital electronics. These peripherals allow for multimodal operation that is essential in a sensing platform. An example of such a device that is coupled with state-of-the-art epidermal electronic form factor is shown in Fig. 10.2B, where NFC is unified with a stretchable antenna that laminates directly to the skin. The device matches the skin in mechanical properties, resulting in seamless integration with the epidermis. Fig. 10.2B illustrates the device's imperceptible nature, showing the high level of conformality illustrated by conformal lamination even under extreme deformation [9]. This high stretchability is engineered by serpentine structures that are arranged in a mutual mechanical plane design. This design features a polyimide (PI) carrier that sandwiches a conductive metal layer, often copper, with another PI layer enabling the structure to deform out of plane in a fashion that limits the strain in the metal layer when stretched. This low strain in the metal layer allows for repeated cycles of strain without conductivity loss. The devices are engineered to withstand global strains of 30%, which is higher than the mechanical compliance of the skin [10,11]. This is visually

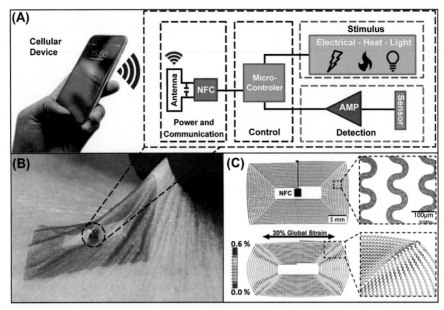

FIGURE 10.2

Schematic overview of a wireless, battery-free epidermal device with the integration of NFC technology. (A) Electronic building blocks of a passive system including NFC, microcontroller to control timed stimulus and recording sequences. (B) Example of an epidermal embodiment with a seamless integration of antenna and SOC in a conformal platform that deforms with the skin [12] © 2014 Wiley-VCH Verlag GmbH & Co. KGaA, Weinheim. (C) Photographic image of stretchable NFC system with corresponding finite element analysis shows the antenna deformed at 30% global strain [12]. 2014 Wiley-VCH Verlag GmbH & Co. KGaA, Weinheim.

documented in Fig. 10.2B and C where the device is stretched on an artificial skin made of silicon elastomer to 30%, and the finite element analysis simulation reveals that the strain in the metal layer is below 0.4% [12], which is well below the yield strain of metal placing the operational strains within the elastic regime. This epidermal form factor enables intimate contact to the skin while weighing only tenths of a gram. This provides an excellent sensor platform that is free from motion artifacts associated with heavy external power supplies or loose coupling with the epidermis common in other commercial devices [13].

Additional benefits of such a platform include the mechanical and chemical robustness against daily wear, such as washing with soap, immersion in water, and sweating during vigorous exercise that would affect normal devices have shown no effect on sensor efficiency [8]. With careful considerations in the type of activity the user will perform while the device is worn; material, electronic, and mechanical properties can be modified and be combined with miniature and wireless electronics to allow for a robust device capable of recording physical health parameters over long periods of uninterrupted communication.

3. NFC-enabled battery-free wireless epidermal/miniature optoelectronic devices

Measuring vitals such as body temperature, heart rate, and peripheral capillary oxygen saturation (SpO2) is important for the overall monitoring of personal health. Heart disease is one of the leading causes of death, accounting for every third fatality in the United States [14]. With the ability to continuously monitor real-time vitals, using wireless, battery-free, optoelectronic devices, vital signs can be closely monitored to introduce preventative measures that result in better treatment outcomes for patients [15].

An example for such a highly integrated device that allows for the wireless collection of clinical grade data streams is shown in Fig. 10.3A [16]. The device is based on a spectrometric principle where small outline LED's are used in conjunction with a photodetector to obtain absorption measurements of the underlying epidermis or object [17]. The system uses four light sources that are equidistant from the detector to measure absorption at four discrete wavelengths [16]. The sequential activation of the light sources is achieved using analog, astable oscillators that are made up of discrete components powered by the NFC system on a chip. The signal at the photodiode is amplified and digitized through an ADC, as shown in Fig. 10.3B.

The resulting embodiment is highly stretchable and allows for conformal epidermal contact. The ultrasoft silicone carrier is doped with a highly absorbent material [18] to improve optical performance by exclusion of ambient light, a photographic image of this black absorbent material and underlying electrical circuit is shown in Fig. 10.3C.

Use of a simple embodiment of the device with only a single infrared (IR) LED (950 nm, AlGaAs), can be used for photoplethysmography, where the arterial pulse waves modulate the absorption of the deoxygenated state of hemoglobin (Hb) and corresponding changes in optical property that can be monitored through a silicon photodetector [19]. The backscattered light allows for measurement of heart rate. With proper calibration of the device, measurements of both systolic and diastolic pressure can be obtained, which can give insight on mean arterial pressure, an indicator of perfusion to vital organs Fig. 10.3D [20,21]. This information can be relayed to a smartphone device to allow for real-time characterization of vitals that can be used in both hospitalization and at home care.

With the addition of a red LED (625 nm), information of both oxygenated and deoxygenated hemoglobin can be continually acquired allowing for the calculation of oxygen saturation. At wavelengths specific to each state of hemoglobin, the Lambert−Beer equations can be used to track changes in concentration resulting

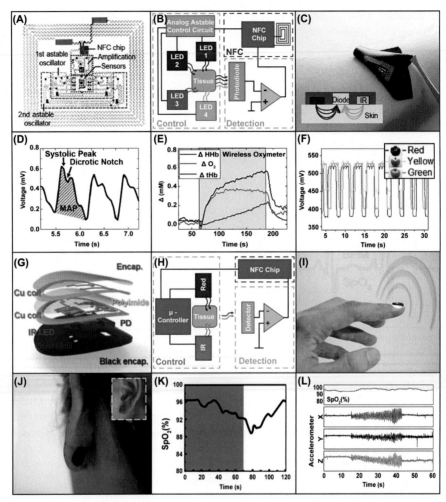

FIGURE 10.3

Wireless and battery-free devices featuring active optoelectronic schemes for the detection of vital signs. (A) Photograph of a wireless, battery-free epidermal spectrometric device including four pulsed LEDs (red, IR, orange, yellow) [16]. (B) Corresponding electrical schematic, which includes an NFC SOC, analog astable control circuit, LED, photodiode, and amplifier. (C) Application example of an epidermal optoelectronic device encapsulated in an opaque stretchable low-modulus silicone elastomer [16]. (D) Wireless measurements using a single wavelength emission to measure heart rate [16]. (E) Measurements of changing deoxyhemoglobin (Hb), oxyhemoglobin (HbO$_2$), and corresponding oxygenation [16]. (F) Color detection of apples using the spectrometric device operated at three discrete wavelengths [16]. (G) Exploded view of the NFC miniature pulse oximeter [19] © 2016 WILEY-VCH Verlag GmbH & Co. KGaA, Weinheim. (H) Block diagram of the wireless pulse oximeter circuit that uses infrared and red LEDs to in the measurements shown in Fig. 10.3E. The absorption of oxygenated states of hemoglobin will change the absorption properties at the red and IR wavelength. By monitoring the backscattered light at two of these wavelengths, controlled by the astable multivibrator analog circuit, the measurement of blood oxygenation can be calculated through calibration of the device with an oxygen—hemoglobin dissociation curve [19,22,23].

With the addition of a yellow and orange LED, a new spectrometric platform can be implemented that allows for color measurement shown in Fig. 10.3F, which is relevant in commercial applications such as the evaluation of food products or clinical application such as measurements that track the change in skin tone over time to help monitor subdermal changes [24,25]. Notable examples of such a disease include jaundice, a disease where bilirubin accumulates in high concentrations in the blood causing the skin to turn yellow, and other defects in iron metabolism causing bronzing of the skin [26,27].

Although offering superior signal-to-noise ratio (SNR) through intimate contact, a limitation of the epidermal platform, however, is the renewal of the epidermis that physically limits wear time to 2—3 weeks depending on the subject. Platforms that are attached to the fingernail allow for a ridged attachment point to extend wearability from weeks to months, with an average growth rate of a fingernail allowing a functional period of 3 months for an adult [19,28].

Using the fingernail as a mounting location requires the device to be significantly miniaturized. In comparison to the epidermal device described earlier, the area of the coil is required to be 13 times smaller for fingernail-mounted devices [16,19]. This can be achieved with advanced wireless coil designs that ensure harvesting and communication capabilities that match that of larger devices. A dual coil design providing the sufficient inductance required for best energy transfer capabilities for these devices is shown in Fig. 10.3G. An absorption measurement scheme analog to the previously described epidermal device can be employed [16], and the electrical operation principle is shown in Fig. 10.3H. Key differences include the precise sequential activation of the IR and red LED light sources using a small microcontroller and corresponding detection of the backscattered signal from the nailbed [29]. The captured time-dynamic spectral information in blood located in the underlying capillary beds allow for the assessment of cardiovascular health [30—32]. The resulting device is shown in Fig. 10.3I. Soft, biocompatible materials incorporated within these devices extend the possible mounting locations on the body to areas such as the ear shown in Fig. 10.3J.

measure hemoglobin species in the nailbed. (I and J) Photographs of the device mounted on both hard and soft surfaces [19,34] © 2016 WILEY-VCH Verlag GmbH & Co. KGaA, Weinheim. (K) Real-time measurements of oxygen saturation during breath hold experiment [19] © 2016 WILEY-VCH Verlag GmbH & Co. KGaA, Weinheim. (L) Stable SpO2 recordings during high levels of motion [19] © 2016 WILEY-VCH Verlag GmbH & Co. KGaA, Weinheim.

The high-fidelity recording capabilities are demonstrated with a breath holding test indicated by the gray area in Fig. 10.3K, where test subjects lower the arterial SpO2 through limited oxygen intake. The continuous measurements of oxygenation reveal a drop corresponding to the limited oxygen intake in the subjects. Noteworthy is the sampling rates, which are higher than traditional counterparts due to the elimination of ambient light and motion artifacts resulting in 2%−3% accuracy for SpO2 estimation and lower requirements for data processing. The sampling rate enables tracking of rapid changes in physiology, which is equivalent to many devices currently used in hospitals [33]. These devices also perform well during high levels of activity showing remarkably stabile and accurate recordings of blood oxygen levels shown in the graph in Fig. 10.3L, where oxygenation is recorded during high acceleration as indicated by simultaneous accelerometer measurements. This dramatic reduction in motion artifacts is attributed to the low mass (51 mg) [19] of the device, a reduction inherent with highly miniaturized and battery-free embodiments.

4. Battery-free and continuous dosimeter applications

There are many areas of human health that are affected by the exposure through ultraviolet (UV) radiation. One area of interest is skin cancer, as it is one of the most prevalent forms of cancer in the United States, killing around 9000 people every year [35]. One of the main risk factors for patients developing malignant skin cancer is the lack of public awareness regarding overall UV radiation exposure [36]. Currently, there are limited options for continuous, real-time monitoring of UV exposure for daily use. This problem motivates the development of wireless, battery-free dosimeter devices that are capable of measuring UV radiation in real time without the need of recharging. Highly miniaturized devices offer an attractive form factor and wearability that is suitable for daily activities [37].

Although the miniaturized embodiments described earlier are reliant on NFC compatible devices located nearby, dosimetry applications can be realized using temporary energy storage such as capacitors that are charged using the sensor itself. This results in autonomous devices that only use NFC-enabled hosts for the readout of the sensor. An exploded view schematic of such a device can be seen in Fig. 10.4A. Charge management of the capacitor, which accumulates charge over the course of a measurement, is realized during the readout process where the system gets discharged during reading events to prevent nonlinearities as outlined in the schematic shown in Fig. 10.4B. With this highly compact system, form factors of 8 mm diameter are possible, resulting in outlines small enough to be mounted permanently on a fingernail shown in Fig. 10.4C. Alternate mounting locations include items such as sunglasses, accessories, and clothing [37]. Epidermally attached devices utilizing the same technology can be worn on multiple skin locations, allowing for a highly accurate measurement of total dosage on the body during daily activities [38].

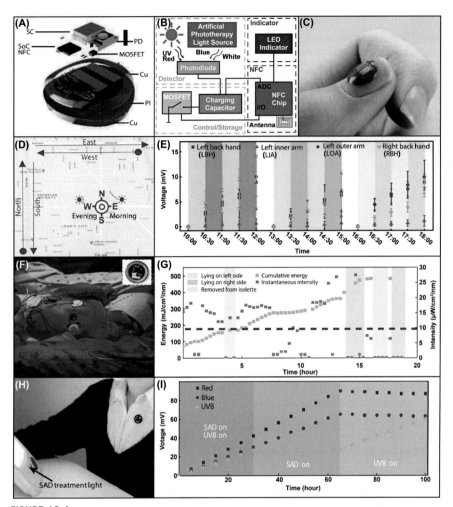

FIGURE 10.4

System overview of a wireless, battery-free, flexible, miniaturized dosimeter to monitor exposure to ultraviolet radiation and light from phototherapy sessions. (A) An exploded view of layered makeup of a battery-free and autonomous UV dosimeter. (B) The PD will convert radiative energy to current which will then be stored on a capacitor that is discharged and recorded by the NFC chip. Data and visible indications can also be implemented. (C) Photographic of a flexible dosimeter that can conform on various parts of the body such as the fingernail. (D) A map showing participants pathways during UV measurement trails. (E) Participant exposure levels are graphed using the mean dosimetry measurements collected on different parts of the body. (F) Image of the device recording dosimetry measurements of blue light phototherapy in the NICU. (G) Measurements of real-time intensity and cumulative dosimetry of a jaundiced infant over the course of 20 h. (H) Image of a wireless dosimeter operating at UVB, red, and blue

An advantage of miniaturized devices is their enhanced, localized sensing ability. Acquisition from multiple points on the body allow for a better understanding of exposure patterns during daily activity as the relative angle, intensity, and exposure time of the affected body part can vary during the day. A typical test used to evaluate the sensing technology over the course of a day is outlined in Fig. 10.4D, where participants walk a predetermined route. The data curated by sensors mounted on various body locations, shown in Fig. 10.4E, show the overall exposure to UV that each participant was exposed to at different times in the day. The correlation of the increase in exposure followed the zenith angles of the sun, which moves during morning, afternoon, and evening. These time periods correlate to zenith angle changes of 65°E to 40°E, 30°E to 20°E, and 30°W to 50°W, respectively. The dosimetry measurements from these participants reveal that specific locations may be significantly more susceptible to UVA exposure compared to other parts of the body [39], such as the inner left arm shown in Fig. 10.4E. Using this new sensing modality results in unparalleled ease of use compared to currently available technology. This allows for access to larger test populations to study UV exposure that may reveal behavioral patterns that can be used as preventative measures for skin disease [40,41]. In a commercial application, real-time readout with a smartphone enables users to monitor and avoid UV overexposure.

This new technology does not only have impact on UV exposure measurements, but also has capabilities to inform clinicians on therapeutics [42]. Using the flexible, battery-free dosimeters, highlighted in Fig. 10.4F, it is possible to study phototherapy in neonates suffering from jaundice [37]. When jaundice is left untreated, bilirubin can accumulate in the brain and cause kernicterus, a form of brain damage that can cause the neonate to become lethargic, deaf, and even increase the probability of developing seizures [43,44]. The resulting information collected from these flexible dosimetry devices allows clinicians to optimize treatment dosage and reduce time spent in the neonatal intensive care unit (NICU). Developing protocols for phototherapy of jaundiced patients in this manner reduces the need of frequent blood analysis, which in newborns, is undertaken via foot pricking that increases the risk of infections and can leave lifelong scars [45,46]. Sensor readouts shown in Fig. 10.4G reveal that an accurate, accumulative dosage can be monitored. This results in individualized recordings allowing for shorter hospitalization times, yielding

---

wavelengths worn on the shirt during white light phototherapy lamp to treat seasonal affective disorder. (I) Continual wireless measurements of phototherapy with combinations of SAD and/or UVB phototherapy sessions.

(A) From Ref. [37]. Reprinted with permission from AAAS. (C) From Ref. [37]. Reprinted with permission from AAAS. (D) From Ref. [37]. Reprinted with permission from AAAS. (E) From Ref. [37]. Reprinted with permission from AAAS. (F) From Ref. [37]. Reprinted with permission from AAAS. (G) From Ref. [37]. Reprinted with permission from AAAS. (H) From Ref. [37]. Reprinted with permission from AAAS. (I) From Ref. [37]. Reprinted with permission from AAAS.

economic benefits, and giving the child more time with their parents, which is crucial in the early days of neonatal development.

Another area of interest for the miniature dosimetry system is to better understand light therapy for conditions such as seasonal affective disorder (SAD) [47]. With an increase in sedentary lifestyles, many lack access to natural outdoor lighting, which can cause many problems in mental health and difficulty sleeping [48,49]. By using a miniature dosimeter like the one shown in Fig. 10.4H, measurements of exposure to UVB and SAD treatment lamps can be achieved to better quantify overall treatment. Here, the wavelength selectivity of the demonstrated device allows for an accurate measurement of accumulated dose as shown in Fig. 10.4I. Wireless, battery-free, flexible, miniaturized dosimeters devices show versatility in multiple applications for consumers and healthcare providers to quantify light exposure across the visible and UV spectrum. These applications aim to reduce risks of excessive exposure to dangerous wavelengths of light and optimize benefits of phototherapies. Physical characteristics of these devices combined with their ability to be completely waterproof, robust, smartphone compatible, and tunable to a wide range of target wavelengths, make these devices important research tools and commercial devices to promote healthier behavioral patterns and study the effects of emerging phototherapies.

5. Miniaturized photo- and electrochemical sensors

Current research in the field of soft, wearable biosensors has been focused on the physical characteristics of the human body such as temperature, blood oxygenation, pedometry, and biopotentials [50,51]. Although these measurements are useful for monitoring many parameters that can hold valuable insight into physiologically relevant processes, they are limited in their medical reach and scope with respect to their monitoring and diagnostic capabilities. Additional modalities are required for the detection and quantification of clinically relevant biomarkers, which can expand the relevance of these systems. Real-time detection of biomarkers is traditionally largely invasive and requires extensive equipment that is rigid, bulky, and inhibitory to everyday activities [52].

Many of these biomarkers are present in blood and other internal fluids; however, to provide patients with a noninvasive solution, targeting external biofluids such as sweat and saliva as a means of monitoring processes inside the body is needed [53]. Sweat has become an increasingly popular biofluid because devices used to collect and analyze it have been constructed in a noninvasive fashion with flexibility of sensing location [38]. Recent studies have shown that sweat contains many biomarkers that, with suitable chemical analysis, hold valuable information regarding metabolic state, hydration, glucose levels, and other important health characteristics [54−56]. Commercial methods of sweat analysis, such the Macroduct, require extensive external hardware and are limited in their ability to detect and analyze multiple different aspects of sweat. Microfluidics modified for the use directly on the epidermis, also known as epifluidics, provide a favorable method of collecting these biomarkers while still operating with a small, easily attachable and chronically wearable platform. An additional advantage over commercially available methods is the lower net volume requirement of biofluid to complete an analysis through the effective use of microfluidic technology. The epifluidic device described in this chapter is only a few centimeters in diameter and can gather and analyze data regarding total sweat loss, sweat rate, pH, as well as glucose, chloride, and lactate concentrations (see Fig. 10.5A) [57]. The epifluidic platform, in conjunction with NFC circuitry, provides a miniaturized, battery-free, noninvasive design for real-time monitoring of biomarkers of interest in sweat and provides a favorable alternative to conventional means of sweat collection and analysis.

The platform utilizes two primary modules for its sensing capabilities: a disposable epifluidic patch for collection and analysis of the sweat, and a reusable electronic module for amplification and digitalization of the electrochemical sensors coupled with an NFC system to relay the data. An exploded view schematic of the layered system is shown in Fig. 10.5A. The epifluidic module consists of a patterned adhesive layer that attaches to the epidermis and controls the inlet location of sweat into the device. The epifluidic layers are made of a silicone elastomer that has been patterned with colorimetric and electrochemical sensing channels with a series of ratcheted channels for gauging sweat loss, and capillary bursting valves (CBVs) for facilitating sweat flow through the device and for allowing time-sequenced sweat sampling. These CBVs are integral to the function of the device. By controlling channel width and diverging outlet angle, the pressure needed to open the valve can be mediated. These valves are engineered such that they are active in the physiological pressure range of sweat and open sequentially. Reagents, dyes, and sensors are fully embedded into the system, providing a smaller net profile of these devices. Material options for these devices are dependent on application and can range from standard polydimethylsiloxane elastomer for regular sportive activity to poly(styrene-isoprene-styrene) for aquatic applications [58]. Device makeup, regardless of material choice, relies of micro-sized fluidic channels as shown in Fig. 10.5B.

The electronic module features a dual layer Pyralux substrate consisting of two 18 μm copper layers separated by a 75 μm, dielectric polyimide layer. This substrate allows for efficient production of thin bilayer PCBs that holds electrical components for data measurement and NFC. This layer is embedded in PDMS for protection from biofluids. This allows for reusability of the more expensive electronic module and the usage of one electronic module with multiple, one use epifluidic devices.

Sweat rate and sweat loss are important factors to consider when determining hydration levels, evaluating autonomic regulation disorders, and identifying stroke [59,60]. Sweat rate is dependent on many factors, including location on the body; however, studies have shown that local sweat loss can be directly correlated to total body sweat loss [51]. To define this metric, ratcheted channels are used to help visualize the flow rate of sweat through the device as shown in Fig. 10.5C. This geometry slows the progression of sweat through the device to prolong its functional analysis

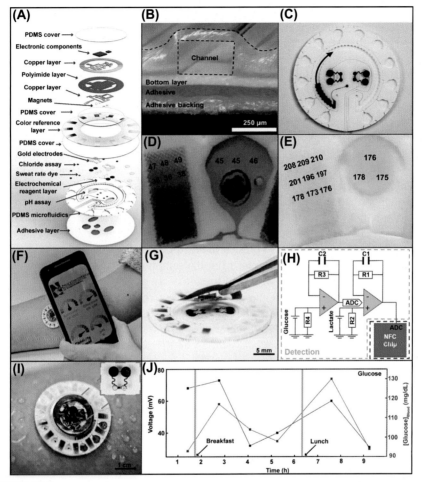

FIGURE 10.5

Schematics and illustrations of complete epifluidic device, sensing modalities, and electrochemical sensing results. (A) Exploded schematic showing the complete composition of the device with epifluidic and electronic systems integrated [57]. (B) Micrograph image of microfluidic channels in an epidermal device [56]. (C) Filling of the sweat rate channel with an arrow indicating the direction of flow [57]. (D) Chrono-sampling well for chloride concentration assay. Colorimetric reference bars shown to the left of the well. Numbers indicate lightness values [57]. (E) Chrono-sampling well for pH assay. Colorimetric reference bars shown to the left of the well. Numbers represent RGB "R" values [57]. (F) Example digital display of sweat composition data being collected and analyzed via NFC and image processing of the user's smartphone [57]. (G) Integration of the electronic platform to the epifluidic system using magnetic attachment [57]. (H) Simplified schematic of the amplification stage used to detect lactate and glucose concentrations [57]. (I) Image showing complete device integration during a sweating. Electrochemical sensors are positioned as indicated by the green markers [57]. (J) Data acquired from biofuel cell-based glucose sensor (black) compared to blood glucose concentrations (red) over a period of 9 h [57].

period. Because sweat rate will vary depending on the application location, altering the inlet dimensions of the adhesive layer enables calibration to different positions on the body. The design shown features a channel that can monitor sweat loss for up to 6 h based on an average sweat rate of $12-120$ μL/h cm^2. A water-soluble dye located near the sweat inlet dissolves when it encounters biofluid, creating a visible fluid that is easily tracked through the channel. This dye not only allow for direct visualization of approximate sweat rate and total sweat loss, but by utilizing a smartphone camera, precise data can be extracted. Image processing techniques available using smartphone cameras enable accurate tracking of the dyed fluid front through the device and can be used to measure sweat rate and total sweat loss. The smartphone application uses computational algorithms, specific to this device, to dissect a collected image and relate changes in optical parameters to meaningful data regarding sweat rate and total sweat loss.

Although sweat sampling is a continuous and irreversible process, time-dependent changes in biofluid composition can be measured through a sequential sampling method and the CBVs present in the microfluidic design. The CBVs allow for time-dependent release of sweat from the sampling channel into testing reservoirs that have been prefilled with colorimetric assay wells. One such assay utilized in this device is the detection and quantification of chloride ions. Chloride ions in sweat can hold great insight into several medical conditions including electrolyte balance and hydration [61]. High chloride concentration in sweat has also been used as a screening test for patients with cystic fibrosis [62]. Fig. 10.5D shows an example chrono-sampling well used for time-dependent detection of chloride in sweat. These assay wells are located at the distal edges of the device with inlets separate from the sweat rate channel. The colorimetric assay utilizes silver chlorinate, a chemical that interacts with chloride ions to produce a purple color, as the catalyst for detection. The silver chlorinate is immobilized in a polyhydroxyethylmethacrylate hydrogel, which allows diffusion of smaller chloride ions into the gel while keeping the concentration of silver chlorinate constant. Changes in lightness, a color and shade independent colorimetric parameter, indicates the concentration of chloride in the sweat sample through a linear calibration curve. Reference bars placed adjacent to the well, allow for ambient light independent concentration readout via smartphone.

Similarly to chloride concentration, pH is another physiologically relevant parameter that can be measured using colorimetric analysis of sweat. Under normal homeostatic conditions, the pH of human sweat falls between 4.5 and 7.0 [50]. In cases of dehydration, electrolyte imbalance, disease, and even stress; the pH of sweat can fluctuate dramatically. Fig. 10.5E shows an example assay used in this device for the detection of pH. Paper pads coated with a pH sensitive dye and a compound to help solubilize ions into the organic phase, known as a phase-transfer catalyst, are placed into the assay well. The paper indicator is cut into the same size as the wells and uses capillary forces to bring sweat into the cellulose membrane. The pH dye ranges from a bright yellow to a deep green color, each with distinct RGB values relating to the pH of the solution. The RBG colorimetric properties of the dye in different pH environments can be compared to a color reference bar to enable smartphone readout of pH.

Use of the reference bars not only allow for quick visual updates on chloride and pH levels, but by using smartphone cameras and image processing techniques, accurate extraction of data regarding chloride concentration and pH can also be acquired. Image processing of these assays take three random points from the test well and measures the colorimetric property appropriate for the specific assay. The mean value from these three points is then compared to the standardized linear curve to yield the chloride concentration and pH. This value is then graphically displayed through an application that is easily accessible for patient or physician use. Fig. 10.5F shows an example of the graphical interface displaying colorimetric and electrochemical analysis sweat composition.

To allow for detachable junction of the modules, each module comes together using a set of thin magnets affixed, using conductive adhesives, to contact pads on the bottom copper layer of the electronics. Another set of magnets are imbedded in the microfluidics platform beneath the electrochemical sensors to enable conductive attachment of the two modules as shown in Fig. 10.5G. The magnets exhibit an attractive force that can hold the modules together through moderate exercise and can be modified to accommodate varying levels of activity.

Glucose and lactate are other biomolecules found in sweat and hold insight into medically important physiological processes. Studies suggest that lactate found in sweat can be used as an indicator of metabolic state and local physical stress [63,64]. A biofuel cell design is utilized in this device as the detection method for this biomarker. The sensor consists of an anode made of carbon nanotube (CNT) paper that allows for conjugation of lactate oxidase, an enzyme that oxidizes lactate to form pyruvate. Tetrathiafulvalene (TTF) is also added to help facilitate electron transport from the enzymes active site to the conductive CNT layer. The anode is coated with a chitosan and a polyvinyl chloride membrane layer to prevent unwanted enzyme leeching and to slow diffusion of lactate to the sensor, thus extending its linear detection range. The cathode consists of gold, layered with platinum black and covered with a Nafion polymer. Oxidation of lactate at the anode and reduction of oxygen to water at the cathode generates an electrical current that is proportional to the concentration of lactate in solution. This current flows across a resistor to generate a voltage that is then amplified and detected as shown in the simplified schematic of Fig. 10.5H. This amplification scheme uses a voltage follower with high-frequency filter to remove noise from the NFC antenna. The amplified signal is then sent to a 14-bit analog-to-digital converter that can transmit information via NFC to a smartphone for real-time analysis as shown before in Fig. 10.5G.

Glucose levels are particularly important monitoring points for patients suffering from diabetes. Traditionally, monitoring of blood glucose levels is carried out using a glucose meter, which requires the patient to draw a small amount of blood using a lancet and then uses disposable test strips and a device to measure blood glucose concentrations. Although this is generally the most accurate method of determining blood glucose levels, it requires the patient to actively measure their glucose concentrations. To address this issue, the development of methodologies to continuously monitor glucose concentrations has been an area of high research and developmental efforts. Current standards in continuous glucose monitoring use a subcutaneously positioned sensor to measure interstitial glucose levels. The sensor transmits data wirelessly to a monitor that displays continuous data and can enable external insulin pumps on a separate device [65]. Although this device provides great insight into blood glucose concentration without the need to draw blood, it is still invasive and requires large form factor electronics to operate effectively. To determine blood glucose concentrations in an effective and noninvasive manner, recent developments have shown that sweat glucose can serve as a qualitative measure for tracking blood glucose [66].

A detection method identical to the lactate sensor is used to identify glucose concentrations with this device. Here, the anode is also coated with a TTF and CNT layer; however, glucose oxidase (GOx) is directly introduced to the Naifon layer which allows for more direct detection of glucose from the sample. This is because the concentration of glucose found in sweat is significantly lower than that of lactate. Current commercial standards of glucose detection utilize a glucose dehydrogenase and pyrroloquinoline quinone reaction scheme, as it is faster than GOx and does not require oxygen. GOx is used in this application because it is a significantly cheaper alternative and the TTF layer acts an electron mediator, thus eliminating the need for oxygen in the reaction. The cathode consists of gold coated platinized carbon in a Nafion film. The GOx oxidizes glucose to gluconic acid while oxygen is, again, reduced to water at the anode to create a current that is proportional to the detected glucose concentration. This current is than converted to a voltage across the load resistor that can be amplified and measured through the same detection scheme.

Fig. 10.5I shows a complete integrated system worn on the forearm during an exercise session. Human trials were used to assess the performance and accuracy of the device. To examine the quality of the colorimetric assays, six subjects performed 15–20 min of cycling while wearing the device and collected periodic sweat samples for commercial gold standard analysis via high-performance liquid chromatography, calibrated pH probes, and nuclear magnetic resonance (NMR) spectroscopy [67]. Results show that colorimetric assays maintained good performance over the course of the test, while NMR analysis seemed to overestimate analyte concentrations detected by the sensor by nearly a factor of 2. Discrepancies in the results are likely due to the difference in real-time data acquisition of the sensor and the spatially averaged, highly processed measurement process of the NMR. The epifluidic sampling system provides real-time collection and analysis of sweat, something that is not possible with current methods. Instead of gathering results from averaged bulk measurements that are temporally inaccurate, the device demonstrates novel, live acquisition of accurate, actionable data that are more representative of the underlying physiological processes.

Fig. 10.5J illustrates the temporal performance of the glucose biosensor over a period of 9 h and the results correlating its voltage reading to standard blood glucose measurements. During the 9-h trail, subjects performed periodic exercise session on a stationary bike. The subjects began the trial in a fasted state and cycled 30 min before and 30 min after consuming breakfast and lunch. Measurements using the device were recorded (black) along with blood glucose reading from a commercially available glucose meter (red). As expected, blood glucose levels rose after meal consumption, dropped over the course of a few hours, and then rose again after the next meal. A similar trend is observed with recordings from the device. Differences in trends can be attributed to contamination from previous sweating periods, as seen in the beginning of the study, and lag time between blood glucose concentration and sweat glucose concentration.

Sweat glucose concentrations lag behind measured blood glucose levels by approximately 30 min to an hour. These lag times are in accordance with previous studies of noninvasive and minimally invasive chemical sensors and can be attributed to biological processes such as capillary exchange [68]. Similar studies conducted for the lactate meter show that the lag time for lactate was significantly smaller. Additionally, these findings agree with previously reported literature aimed at defining the relationship between sweat glucose and lactate levels collected and analyzed ex situ, with concentrations present in blood [66,69].

Collection and analysis of sweat provides a noninvasive means of detecting many physiological underpinnings that would otherwise require invasive equipment that would hinder the quality of life of the patient. Current means of sweat biomarker detection does not satisfy these goals and relies heavily on external hardware. NFC electronics in conjunction with epifluidic technology provides a robust means of multimodal detection on a small platform that can full integrate onto the epidermis for sweat collection and analysis without the need for extensive, external hardware.

---

6. Thermal sensors

The optical and electrochemical detection modes described earlier already offer a broad set of physical detection capabilities; however, quantitative evaluation of thermoregulatory function, hydration, blood perfusion, wound healing, and other health criterions beyond subdermal depths are difficult or impossible to realize. To overcome these barriers, the use of thermal actuators and sensors allows for sensing capabilities of these parameters that penetrate beyond the epidermis. Sensors and actuators with conformal contact with the epidermis allow for greater thermal coupling and thus greater resolution as a result from the high signal-to-noise performance of the sensing element. Current measurement devices are reliant on electrical impedance measurements that are limited by the requirement of trained medical professionals to perform placement, and the need for precise pressure calibration at the contacting interface via handheld probes [70,71]. Here, sensing methods derived from materials testing offer a solution via thermal actuation and simultaneous sensing, which can produce precise time-dependent data that can be analyzed to determine thermal conductivity and thermal diffusivity that can be related to detailed information on blood flow, skin hydration, wound healing, sunburn, and cellulitis [4,72–75]. The NFC miniaturized electronics described earlier coupled with thin-film heaters that have epidermal contact, can be realized to allow for wireless measurements of thermal properties at various locations on the body that extend several millimeters under the skin [76]. Such epidermal wireless thermal sensors (eWTS) can operate battery free with data streams extending multiple days.

A schematic illustration of an eWTS appears in Fig. 10.6A and consists of two mechanically distinct components. The first involves an NFC-based flexible device with similar components to the systems described in the previous section and analog signal conditioning that involves an active H-bridge to drive the sensor actuator combination. The second component consists of a photolithographically patterned thin thermal sensor on an elastomeric substrate with a clearly defined temperature coefficient of resistance (TCR). An opening in the flex-PCB serves as a location for integrating the epidermal platform via an acrylate-based, pressure sensitive label adhesive. A thin medical grade adhesive allows for reversible bonding of the flex-PCB to the skin.

A microcontroller enables precise timing and control over the actuation of the eWTS, and an analog driver provides a constant heating power that results in a resistance change amplified with onboard operational amplifier. This system allows for high resolution, wireless measurement of changes in temperature [73]. A schematic of the circuit is shown in Fig. 10.6B. The resistive, thin-film metallic element ($S1 \approx 650 \, \Omega$) located on one arm of the Wheatstone bridge (R2, R3, R4) enables the circuit's sensing capabilities. The TCR of the metallic element determines the relationship between changes in resistance and temperature. Changes in S1 resistance cause a differential voltage change, which is amplified and converted into a continuous data stream. To induce thermal actuation, the microcontroller controls the S1 timing. To ensure robust operation in an RF field, this signal is buffered by an operational amplifier in a voltage-follower configuration that is located on the same die as the operational amplifier. The amplified signal is fed into the analog-to-digital converter of a bare-die NFC chip that features a 10-bit ADC with a dynamic range of 300 mV, resulting in a resolution of 80 mK [77].

High stretchability of the device is derived from a mutual mechanical plane design configuration which features a symmetric polyimide encapsulation of the gold thin film conductor. Fig. 10.6C shows this high mechanical compliance where a cotton-tipped applicator easily deforms the soft, stretchable device. The low rigidity and low modulus facilitate conformal contact with the skin [78]. The capacity for these sensors to form strong, reversible contact between the skin and actuator is a critical design consideration. The overall size ($D = 1.6 \, \text{cm}$) and low mass (200 mg) allow for application across a broad range of body locations [79]. The compact size of the device reduces the energy release rate, which determines delamination, and in turn minimizes the work of adhesion [80]. The polyimide layers mechanically and electrically isolate the thin thermal sensor, resulting in low hysteresis behavior and robust operation of the overall device.

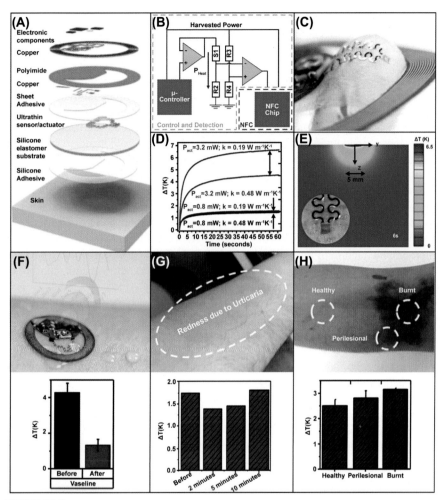

FIGURE 10.6

Device design, requirements, and measurements for thermal characterization on human subjects. (A) Exploded view schematic illustration of an epidermal wireless thermal sensor (eWTS) [77] © 2018 WILEY-VCH Verlag GmbH & Co. KGaA, Weinheim. (B) Illustration of the circuit design, showing H-bridge components involved in sensing and control [77] © 2018 WILEY-VCH Verlag GmbH & Co. KGaA, Weinheim. (C) Photographic image of the sensor held with a pair of tweezers [77] © 2018 WILEY-VCH Verlag GmbH & Co. KGaA, Weinheim. (D) Transient temperature response for two different actuation powers ($P_{act} = 3.2$ and 0.8 mW), on two skin phantoms (Sylgard 184, red and black curves and Sylgard 170, pink and blue curves) acting as examples for skin [77] © 2018 WILEY-VCH Verlag GmbH & Co. KGaA, Weinheim. (E) Side view of temperature distribution computed by 3D-FEA 6 s after actuation at 2 mW and micrographic view of sensor from aforementioned [77] © 2018 WILEY-VCH Verlag GmbH & Co. KGaA, Weinheim.

The high sensitivity of the system can be demonstrated with in situ experiments using commercially available silicone materials mimicking the thermal properties of the skin, as shown in Fig. 10.6D. Temperature sensing combined with low-power actuation results in a recorded rise in temperature. The observed behavior depends mainly on thermal diffusivity of the material immediately following actuation and on thermal conductivity for times beyond 2 s [74]. Thermal conductivity is determined as the inverse of change in temperature at a fixed time interval.

The measuring capabilities of the eWTS platform have been validated by extensive comparisons to wireless data, IR imaging the current gold standard in clinical applications, and finite element analysis (FEA) models [74a]. A 3D-FEA simulation in cross sectional view of temperature distribution 6 s after thermal actuation at 2 mW along with a photographic top view image of the eWTS (inset) is shown in Fig. 10.6E. The simulation provides insight into the characteristic thermal penetration depth of several millimeters into the epidermis with a change in surface temperature of 6 K [73], which is below the sensation threshold for humans [77]. The inset micrographic picture of the eWTS in Fig. 10.6E highlights the polyimide layer surrounding the actuator/sensor and the serpentine connects allowing for conformal contact with the skin.

The small size, stretchable mechanics, lightweight, battery-free operation, and waterproof design of the eWTS platform allow for chronic wearability over a range of conditions for several days. Device lifetimes extended over 1 week in subjects who wore the eWTS and followed normal daily activities including exercise, shower, and sleep. Ambient water content did not affect RF communication or circuit functionality [81], allowing for stable operation as illustrated in Fig. 10.6F.

The human epidermis is a complex network requiring constant hydration. Free water in the outer and inner layers of the human epidermis serves a structural role in maintaining the lamellar arrangement of the lipid matrix, which is vital for preventing transepidermal water loss and absorption of airborne pollutants and pathogens [82]. Moisture levels in the outer layers of the epidermis are ordinarily about 20%. Levels below 10% result in xeroderma, characterized by poor barrier

◀─────────────────────────────────────

(F) Photographic image of wet eWTS, simulating sweating conditions with graph showing transient temperature response of skin before and 15 min after application of Vaseline using eWTS platform. Standard deviations computed across three measurements provided error bars [77,85] © 2018 WILEY-VCH Verlag GmbH & Co. KGaA, Weinheim. (G) Photographic image of a subject's arm 2 min after dermatographic urticaria, showing inflammation and redness with graph showing measured temperature change 6 s after actuation before, 2, 5, and 10 min after urticaria [77] © 2018 WILEY-VCH Verlag GmbH & Co. KGaA, Weinheim. (H) Photographic image of subject's arm 2 days after suffering burns. Burnt, healthy, and perilesional skin are displayed with correlated measured temperature changes 6 s after actuation at three locations of the subject's arm [77] © 2018 WILEY-VCH Verlag GmbH & Co. KGaA, Weinheim.

properties and the appearance of dry flaky skin [82]. Hence, the measurement of epidermal moisture levels is important for the fields of dermatology, toxicology, and sports medicine.

The eWTS platform for skin hydration characterization can be compared to the gold standard, which is an impedance-based measurement. Previous work shows that thermal conductivity of the skin correlates linearly to its hydration [4,73]. Here, eWTS devices assess thermal conductivity to detect changes in the filling fraction of water in the collagen matrix. An experiment involving the measurement of the transient temperature response of skin applied with Vaseline, a humectant used to reduce the rate of transdermal water loss, before and 15 min after application to the skin is shown in Fig. 10.6F.

Changes in thermal transport of the skin due to trauma can be studied using eWTS platforms. An example of such capabilities involves slap induced dermatographic urticaria, which results in temporary inflammation and hyperemia, prompting increased rates of thermal transport via an increased volume of blood perfusing the tissue [83]. An image of urticaria is shown in Fig. 10.6G, and corresponding thermal transport properties of the skin over time are shown in Fig. 10.6G. Here, measurements with the eWTS show higher thermal conductivity corresponding to the physiological state of the skin. The measurements reflect the recovered thermal conductivity values returning back to baseline, highlighting the great temporal accuracy of the system even compared to wired embodiments with external data acquisition worth multiple thousands of dollars [4].

Other types of trauma, such as a class II burn injury, lead to ischemia and necrosis of the dermal layers, reduced perfused tissue, and lowered skin conductivity [84]. An image of the right forearm of a burn victim is illustrated in Fig. 10.6H, showing burnt, perilesional, and healthy skin. Measurements in thermal conductivity 2 days after the burn across the three types of skin are displayed in Fig. 10.6H revealing that thermal conductivity decreases with burn damage to the skin, establishing the platform as a great tool to monitor healing progression in burn victims.

The developments presented here exhibit concepts in electrical, mechanical, and materials design as avenues to lightweight, wearable sensors for multimodal thermal characterization with precision equivalent or superior to that of the clinical gold standard. These devices pose utility in monitoring prevalent clinical problems, such as skin hydration, that cannot be addressed with current technologies. By detecting changes in thermal conductivity across the skin, the eWTS platform provides a ubiquitous method for quantifying skin hydration. These platforms can also be easily mounted on different body locations such as the arm, neck, and leg suggesting diverse utility in a range of clinical problems. The sensor design affords pathways for low-cost scaling and manufacturing to meet the requirements of large study populations, with ultimate utility in clinical and at home monitoring performed by ordinary users.

7. Flow sensors

Ventricular shunts serve as a primary treatment for patients with hydrocephalus, a neurological disorder resulting from the overproduction and/or reabsorption of cerebrospinal fluid in the ventricular system of the brain. Current diagnostic tests for shunt function suffer from excessive cost, poor reliability, low speeds, susceptibility to interference, and patient discomfort. Shunted patients undergo an average of two CT scans annually, resulting in dangerous cumulative amounts of radiation exposure that have been linked to the onset of neurological and hematological malignancies [86,87]. Current attempts to assess shunt function include large thermal sensors coupled with an icepack to sample downstream temperature fluctuations; however, application is highly manual and requires trained personnel. Miniaturized, epidermal thermal sensors pose a promising avenue for noninvasive flow sensing to assess shunt function.

Fig. 10.7A shows a device that relies on similar sensing modalities in a drastically different form factor; here, a heater creates the temperature gradient that modulates the temperature at fixed distance to sense flow, flow rate, and flow direction. The device adheres robustly and makes epidermal contact with the skin. As shown in Fig. 10.7B, the epidermal linear array (ELA) consists of an actuator and a pair of sensors located 1.5 mm upstream and downstream of the actuator, respectively. Here, the heater is used to create a difference in upstream and downstream temperature, providing flow characterization by the two sensors. The actuator simultaneously serves as a temperature sensor, and the measured temperature of the actuator is used as a normalizing factor to facilitate data analysis independent of actuation power [88]. When tested on skin phantoms, thermal sensors show that fluid transported heat away from the actuator preferentially to the downstream sensor and away from the upstream sensor, resulting in a thermal anisotropy induced by macrovascular blood flow through the skin.

Fig. 10.7C illustrates recorded wireless flow measurements for two different flow rates as a function of time. During the interval for which flow occurs, the temperature difference (ΔT) between the upstream and downstream sensors sustains a positive value until flow ends. The low levels of ΔT during periods of no flow and high levels of ΔT during the period of flow support the fidelity of the on-skin sensors. Low flow rates produce the greatest difference between upstream and downstream temperatures, as high flow rates lead to convective cooling of the sensors. Flow rate is determined by comparing the temperature difference between the sensors, a measure of thermal anisotropy ($\Delta T_{Sensor}/T_{Actuator}$), and their average, a measure of flow magnitude ($\overline{T}_{Sensor}/T_{Actuator}$).

Increasing the power of the actuator ($P_{actuator}$) improves the SNR of the measurements, but biological considerations such as sensation and pain threshold set the upper limit. Recent work has established a set of design considerations and algorithms for determining depth-dependent thermal properties of soft tissue to depths of up to ~6 mm [72]. Such sensing modalities have clinical application in measuring flow through cerebrospinal shunts wirelessly and in real time to identify shunt

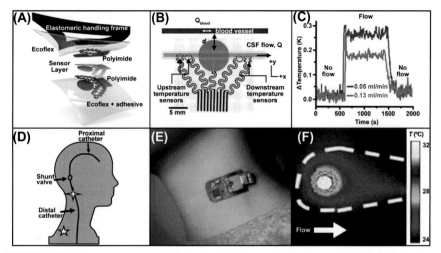

FIGURE 10.7

Wireless flow sensors with patient trials. (A) Exploded view illustration of an ELA to be used in a hospital setting, with elastomeric handling frame, ecoflex, and adhesive. (B) Photographic image with superposed illustration of ELA function on a catheter and a blood vessel. (C) Wireless upstream and downstream temperature difference measurements as a function of time for two different flow rates, Q = 0.05 mL/min and Q = 0.13 mL/min. (D) Depiction of on-shunt (4-point star) and off-shunt (5-point star) ELA placement on a shunted patient. (E) Image of ELA positioning on a shunted patient. (F) IR thermograph highlighting the anisotropy of the distribution of temperature in the presence of flow.

(A) From Ref. [88]. Reprinted with permission from AAAS. (B) From Ref. [88]. Reprinted with permission from AAAS. (C) From Ref. [88]. Reprinted with permission from AAAS. (D) From Ref. [88]. Reprinted with permission from AAAS. (E) From Ref. [88]. Reprinted with permission from AAAS. (F) From Ref. [88]. Reprinted with permission from AAAS.

malfunction or failure. These measurements require placement on the skin over the distal catheter and at a location adjacent to the distal catheter as displayed in Fig. 10.7D. The measurement at the adjacent location serves as a control for comparison with on-shunt measurements. A photographic image of the device placed on a shunted patient is shown in Fig. 10.7E. The soft, removable, skin-safe adhesive with a low thermal mass and conformal contact allows for nonirritating contact, motion artifact free operation, and easy removal. Raw ADC values are converted to temperature using a simple linear calibration for two sensors on the device and relayed wirelessly using highly miniaturized radios to computer and smartphone infrastructure.

Although ELA devices utilize only an upstream and downstream sensor, expanded epidermal sensing arrays (ESAs) for characterizing flow through shunts have incorporated a central thermal actuator connected to 100 precision temperature sensors via serpentine interconnects. The large sensor array facilitates an easy placement on a patient even without high resolution CT images or prior knowledge of exact shunt location. Temperature differentials for each equidistant sensor pair with respect to the thermal actuator determine the extent of thermal anisotropy that results from flow. Linear calibration of the sensor outputs yields high-quality thermal map constructed from ESA data as shown in Fig. 10.7F, where thermal transport is greatest in the direction of flow. A defined anisotropic temperature distribution, with the magnitude reflecting volumetric flow rate, allowes for determination of subcutaneous shunt function via computer interface [88].

The skin-like, precision flow sensor systems presented here have the potential to advance the area of clinical diagnostics of shunt malfunctions and other illnesses. Compared to radiographic imaging, invasive sampling, and ice pack cooling, these platforms are unique in their combination of wireless data transmission and soft, epidermal thermal sensors. Conditions such as normal pressure hydrocephalus and idiopathic intracranial hypertension result from neurological hydrodynamic dysfunction yet lack the tools for precise diagnosis and would benefit from on-skin flow sensors. The devices presented here offer further quantitative modes of use including measuring intracranial pressure and near-surface blood flow. With the implementation of such devices, new and improved treatment approaches can potentially be developed to care for these conditions.

8. Devices on market

Current devices used for detection and recording of biodata are limited in their scope as they lack the necessary properties for comprehensive, minimally invasive data acquisition and analysis. For these devices to be viable candidates for broadly distributed tools that deliver clinical grade information in a form factor that is amendable to continuous wear, the utilization of wireless technology is essential.

The devices presented in this chapter utilize technology that has been demonstrated to provide a wide array of clinically relevant data while using many different sensing modalities. Soft, epidermal, and battery-free systems provide an extremely desirable alternative to conventional practices due to their imperceptible nature. To broadly disseminate these devices, there is a need for companies to adapt this platform to help bring this technology to market and begin integrating these devices into common medical practices. Broad dissemination will help make this technology more commonplace, leading to chronically collected data streams available for analysis through smart algorithms, resulting in advanced diagnostic and preventative digital healthcare.

L'Oréal, a leader in skincare products and technology, has been one of the first companies to adapt technology presented in this chapter to create a commercially available product. Their My Skin Track UV, shown in Fig. 10.8A, utilizes battery-free dosimetry with NFC technology, similar to the device described in the dosimetry section, as a means of tracking environmental factors that are important

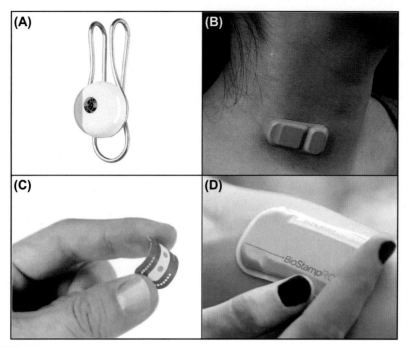

FIGURE 10.8

Overview of current devices utilizing epidermal and NFC device technology that are available on market. (A) L'Oréal's My Skin Track UV device used to track many different environmental agitators. Reproduced with the permission of Wearifi Inc. (B) Rhaeos's wearable shunt monitor for monitoring patient shunt function. Reproduced with the permission of Northwestern University. (C) L'Oréal's My Skin Track pH device used to determine the pH of the user's skin. Reproduced with the permission of Epicore Biosystems. (D) MC10's BioStamp being applied to a patient for continuous activity. Reproduced with the permission of MC10 Inc.

determinants of skin health. The My Skin Track UV is the first wireless, wearable electronic device that measures UVA and UVB exposure and can provide patients with detailed analysis regarding UV radiation exposure. The device is small and unobtrusive being only 12 mm wide and 6 mm tall. It is waterproof and has a wire clip that can be attached to clothing or other personal accessories. The data is collected and sent to an application on a smartphone for instant updates, with the capacity to store up to 3 months of data. Data is analyzed and presented in a way for consumers to make informed choices for their skincare.

Although many patients may use this information for general health and beauty decisions, this device has further implications in the medical field. In the last few years, the rate of malignant skin tumors, or melanoma, has increased as a result of increased and prolonged UV exposure [89]. Many individuals may be unaware of their exposure, the My Skin Track UV can help patients make more informed decisions regarding their total UV exposure and may help decrease the incidence of UV and other environmentally related diseases.

Rhaeos, a company commercializing the work discussed in the flow sensors section, has adapted the epidermal flow technology shown in Fig. 10.8B to develop a commercial device for tracking ventricular shunt function in patients suffering from hydrocephalus. The device, shown in Fig. 10.8B, uses ESA to track and quantify anisotropies in epidermal thermal transfer characteristics induced by macrovascular blood flow. The electronics are embedded in a soft, flexible polymer that can adhere to the skin. ESAs are placed upstream and downstream of the shunt to monitor flow parameters and detect abnormalities.

Ventricular shunts are usually set in place for eight or more years without the need for replacement; however, there are many complications that can occur, and the shunts require frequent monitoring and assessment. Mechanical failure, obstruction, and infections can all cause serious complications in shunt activity and lead to over- and underdrainage of cerebrospinal fluids. These complications can cause headache, brain hemorrhage, and the return of hydrocephalus symptoms. The device is designed to be a noninvasive, low-burden solution that can provide accurate information to patients and healthcare providers. As the technology advances, this system can be adapted to detect a multitude of flow related illnesses at deeper, subdermal distances.

Epicore biosystems is a company focusing on the development of "skin-like" epifluidic devices for the real-time collection and analysis of sweat biomarkers. They have adopted similar technology for sweat collection and analysis as described in the chemical sensor section of this chapter. They have partnered with many companies, including L'Oréal and Gatorade, to help bring this technology to the marketplace. In collaboration with La Roche Posay, they have created the My Skin Track pH. The My Skin Track pH, seen in Fig. 10.8C, is a small device that can detect sweat pH levels and communicate with a smartphone application using image processing. Healthy skin generally exhibits a pH in the rage of 4.5—5.5 and when the pH fluctuates, the skin becomes increasingly susceptible to external factors that can cause several negative effects on beauty and health. These negative effects range from unwanted wrinkles to eczema and atopic dermatitis [90]. The device utilizes soft, epifluidic technology to accurately route sweat through the device to the pH assay and viewing point. After the wearer has worn the sensor for 5—15 min, they use a smartphone with the My Skin Track pH app to image the sensor. The smartphone application processes the data and will give the user information regarding pH, local sweat loss, and sweat rate. This information can be used to help inform users on the health of their skin and products that could help combat their unique condition. Although this technology has been commercialized in this market, sweat-based sensors have extensive need outside of skin care. Athletes performing at the highest level have used such technology to enhance their performance and recovery. Sensors tracking electrolyte composition, like the one described previously, help give athletes informative data on their hydration and how to best supplement nutrition. In a healthcare setting, hydration is a key component of disease recovery and monitoring hydration status of patients can improve outcomes and reduce financial burdens from extended stays or repeated visits. The skin patch technology being developed and commercialized provides the most accessible, widely applicable solution to date.

Another commercially viable example of this epidermal technology is MC10's BioStamp. MC10's focus as a company is developing hardware and software systems to help provide informational healthcare analytics with minimal hindrance of daily activities. Their flagship product is the BioStamp nPoint, shown in Fig. 10.8D. It is a wireless, rechargeable epidermal device that can be utilized by researchers or healthcare professionals to collect and analyze physiological data without the need for direct patient interaction. The device is a soft, flexible patch that contains multimodal sensing capabilities and can store up to 36 h of actionable data such as sleep, posture, activity, and vital signs.

The full system includes the stamps with a charging dock that charges the device and synchronizes data, as well as two user applications. The in-home MC10 link app provides patients with instructions on how to use the device and can deliver alerts and reminders to take medication or perform other duties. The MC10 investigator app provides researchers and clinicians with both raw and derived data that can give them insight into proceedings of clinical trials or information on patient progress. This type of technology platforms enables more effective clinical trials, as shown in a 2017 study [91]. Although this study focused primarily on patients with Parkinson and Huntington diseases and used single modality sensors, results showed that utilization of these sensors were feasible and suitable for patients and researchers. Other ongoing studies that utilize the Biostamp technology include monitoring of postural sway and gait in patients with multiple sclerosis, intraoperative monitoring of neuromuscular function, and long-term vector cardiogram monitoring in patients with heart failure. The multimodality of the BioStamp nPoint can expand the scope of use for this technology to broader applications and provide a versatile platform for further development of soft sensors and digital healthcare.

9. Summary

The current means of recording, analyzing, and storing biodata is outdated and limits data quality, data relevancy, and lacks capabilities to record chronic timeframes relevant to the clinical problem. These shortcomings compromise the accuracy of diagnosis, leading to suboptimal patient outcome. Epidermally attached, soft devices provide a promising alternative to current technology by allowing for fully biointegrated components that can utilize NFC and wireless power harvesting for chronic recording and digital data analysis. Although the aforementioned companies are making strides toward more biointegrated technology in diagnostic devices, the current scope of the technology on the market is still in its infancy. Further proliferation of battery-free imperceptible biosensors coupled with cloud-based data collection and analytics will help to improve the quality of life of many patients and advance medical research.

From optoelectronics and dosimetry to electrochemical and thermal sensing, the systems covered in this chapter have proven their capacity to bring high fidelity, multimodal sensing to many different subfields of healthcare. These devices have shown to be viable alternatives to contemporary counterparts by demonstrating equal or superior sensing capabilities in a miniaturized, battery-free form factor.

Further development of this platform holds the potential to yield multimodal and highly miniaturized devices that serve as imperceptible monitors of health that deliver data streams capable of recording valuable information to deliver early diagnosis and ongoing status during treatment to realize highly personalized medicine. Advancement and refinement of these technologies are essential for paving the way for the fourth industrial revolution, where physical, digital, and biological sciences come together to create biointegrated, digital systems that vastly improve current diagnostic and treatment pathways to achieve a higher standard of healthcare and personal health.

References

[1] Obermeyer Z, Emanuel EJ. Predicting the future — big data, machine learning, and clinical medicine. N Engl J Med 2016;375:1216—9.

[2] Bonato P. Advances in wearable technology and applications in physical medicine and rehabilitation. J Neuroeng Rehabil 2005;2:2.

[3] Tian L, et al. Large-area MRI-compatible epidermal electronic interfaces for prosthetic control and cognitive monitoring. Nat. Biomed. Eng. 2019. https://doi.org/10.1038/s41551-019-0347-x.

[4] Webb RC, et al. Epidermal devices for noninvasive, precise, and continuous mapping of macrovascular and microvascular blood flow. Sci. Adv. 2015;1:e1500701.

[5] Dagdeviren C, et al. Conformal piezoelectric systems for clinical and experimental characterization of soft tissue biomechanics. Nat Mater 2015;14:728.

[6] Want R. Near field communication. IEEE Pervasive Comput 2011;10:4—7.

[7] Kim H-J, et al. Review of near-field wireless power and communication for biomedical applications. IEEE Access 2017.

[8] Han S, et al. Battery-free, wireless sensors for full-body pressure and temperature mapping. Sci Transl Med 2018;10.

[9] Lacour SP, Jones J, Wagner S, Li T, Suo Z. Stretchable interconnects for elastic electronic surfaces. Proc IEEE 2005;93:1459—67.

[10] Silver FH, Freeman JW, DeVore D. Viscoelastic properties of human skin and processed dermis. Skin Res Technol 2001;7:18—23.

[11] Ní Annaidh A, Bruyère K, Destrade M, Gilchrist MD, Otténio M. Characterization of the anisotropic mechanical properties of excised human skin. J Mech Behav Biomed Mater 2012;5:139—48.

[12] Kim J, et al. Epidermal electronics with advanced capabilities in near-field communication. Small 2015;11:906—12.

[13] Motti VG, Caine K. Human factors considerations in the design of wearable devices. Proc Hum Factors Ergon Soc Annu Meet 2014;58:1820−4.

[14] Dariush M, et al. Heart disease and stroke statistics−2015 update. Circulation 2015; 131:e29−322.

[15] Lymberis A. Smart wearable systems for personalised health management: current R&D and future challenges. In: Proceedings of the 25th annual International conference of the IEEE engineering in medicine and biology society (IEEE cat. No.03CH37439), vol. 4; 2003. p. 3716−9. 4.

[16] Kim J, et al. Battery-free, stretchable optoelectronic systems for wireless optical characterization of the skin. Sci Adv 2016;2. e1600418−e1600418.

[17] Lochner CM, Khan Y, Pierre A, Arias AC. All-organic optoelectronic sensor for pulse oximetry. Nat Commun 2014;5:5745.

[18] Jang K-I, et al. Rugged and breathable forms of stretchable electronics with adherent composite substrates for transcutaneous monitoring. Nat Commun 2014;5:4779.

[19] Kim J, et al. Miniaturized battery-free wireless systems for wearable pulse oximetry. Adv Funct Mater 2017;27:1604373.

[20] Kimble KJ, Darnall RAJ, Yelderman M, Ariagno RL, Ream AK. An automated oscillometric technique for estimating mean arterial pressure in critically ill newborns. Anesthesiology 1981;54:423−5.

[21] Yamakoshi K, Shimazu H, Shibata M, Kamiya A. New oscillometric method for indirect measurement of systolic and mean arterial pressure in the human finger. Part 1: model experiment. Med Biol Eng Comput 1982;20:307−13.

[22] Chan ED, Chan MM, Chan MM. Pulse oximetry: understanding its basic principles facilitates appreciation of its limitations. Respir Med 2013;107:789−99.

[23] Collins J-A, Rudenski A, Gibson J, Howard L, O'Driscoll R. Relating oxygen partial pressure, saturation and content: the haemoglobin-oxygen dissociation curve. Breathe 2015;11:194−201.

[24] Gaigalas AK, Wang L, Karpiak V, Zhang Y-Z, Choquette S. Measurement of scattering cross section with a spectrophotometer with an integrating sphere detector. J Res Natl Inst Stand Technol 2012;117:202−15.

[25] Nishino S, Ohshima K. A feasibility study of gender recognition with a near−infrared ray scanning spectrophotometer. Electron Commun Jpn 2011;94.

[26] Porter ML, Dennis BL. Hyperbilirubinemia in the term newborn. Am Fam Physician 2002;65:599−606.

[27] Makker J, Hanif A, Bajantri B, Chilimuri S. Dysmetabolic hyperferritinemia: all iron overload is not hemochromatosis. Case Rep. Gastroenterol. 2015;9:7−14.

[28] Kim J, et al. Miniaturized flexible electronic systems with wireless power and near-field communication capabilities. Adv Funct Mater 2015;25:4761−7.

[29] Mendelson Y, Ochs BD. Noninvasive pulse oximetry utilizing skin reflectance photoplethysmography. IEEE Trans Biomed Eng 1988;35:798−805.

[30] Yokota T, et al. Ultraflexible organic photonic skin. Sci Adv 2016;2:e1501856.

[31] Sinex JE. Pulse oximetry: principles and limitations. Am J Emerg Med 1999;17: 59−67.

[32] Tremper KK. Pulse oximetry. Chest 1989;95:713−5.

[33] Jubran A. Pulse oximetry. Crit Care 1999;3:R11−7.

[34] Kim J, et al. Oximetry: miniaturized battery-free wireless systems for wearable pulse oximetry. Adv Funct Mater 2017;27.

[35] American Cancer Society. Cancer facts and figures 2018. 2018.

[36] Nahar VK, et al. Skin cancer knowledge, attitudes, beliefs, and prevention practices among medical students: a systematic search and literature review. Int J Women's Dermatol 2018;4:139−49.

[37] Heo SY, et al. Wireless, battery-free, flexible, miniaturized dosimeters monitor exposure to solar radiation and to light for phototherapy. Sci Transl Med 2018;10.

[38] Ray TR, et al. Bio-integrated wearable systems: a comprehensive review. Chem Rev 2019. https://doi.org/10.1021/acs.chemrev.8b00573.

[39] Moseley H, et al. Guidelines on the measurement of ultraviolet radiation levels in ultraviolet phototherapy: report issued by the British Association of Dermatologists and British Photodermatology Group 2015. Br J Dermatol 2015;173:333−50.

[40] Imoto K, et al. The total amount of DNA damage determines ultraviolet-radiation-induced cytotoxicity after uniformor localized irradiation of human cells. J Investig Dermatol 2002;119:1177−82.

[41] Masuma R, Kashima S, Kurasaki M, Okuno T. Effects of UV wavelength on cell damages caused by UV irradiation in PC12 cells. J Photochem Photobiol B Biol 2013;125:202−8.

[42] Ferreira GR, de Vasconcelos CKB, Bianchi RF. Design and characterization of a novel indicator dosimeter for blue-light radiation. Med Phys 2009;36:642−4.

[43] Thyagarajan B, Deshpande SS. Cotrimoxazole and neonatal kernicterus: a review. Drug Chem Toxicol 2014;37:121−9.

[44] Bhutani VK, Wong RJ. Bilirubin neurotoxicity in preterm infants: risk and prevention. J Clin Neonatol 2013;2:61−9.

[45] Onesimo R, et al. Is heel prick as safe as we think? BMJ Case Rep 2011;2011. bcr0820114677.

[46] Koklu E, Ariguloglu EA, Koklu S. Foot skin ischemic necrosis following heel prick in a newborn. Case Rep Pediatr 2013;2013:912876.

[47] Roecklein KA, Rohan KJ. Seasonal affective disorder: an overview and update. Psychiatry 2005;2:20−6.

[48] Owen N, Sparling PB, Healy GN, Dunstan DW, Matthews CE. Sedentary behavior: emerging evidence for a new health risk. Mayo Clin Proc 2010;85:1138−41.

[49] Collings PJ, et al. Prospective associations between sedentary time, sleep duration and adiposity in adolescents. Sleep Med 2015;16:717−22.

[50] Wang S, Zhang G, Meng H, Li L. Effect of exercise-induced sweating on facial sebum, stratum corneum hydration, and skin surface pH in normal population. Skin Res Technol 2013;19:e312−7.

[51] Choi J, Ghaffari R, Baker LB, Rogers JA. Skin-interfaced systems for sweat collection and analytics. Sci Adv 2018;4:eaar3921.

[52] Kim SB, et al. Soft, skin-interfaced microfluidic systems with wireless, battery-free electronics for digital, real-time tracking of sweat loss and electrolyte composition. Small 2018;14:1802876.

[53] Bandodkar AJ, Wang. J. Non-invasive wearable electrochemical sensors: a review. Trends Biotechnol 2014;32:363−71.

[54] Gao W, et al. Fully integrated wearable sensor arrays for multiplexed in situ perspiration analysis. Nature 2016;529:509.

[55] Sonner Z, et al. The microfluidics of the eccrine sweat gland, including biomarker partitioning, transport, and biosensing implications. Biomicrofluidics 2015;9:31301.

[56] Koh A, et al. A soft, wearable microfluidic device for the capture, storage, and colorimetric sensing of sweat. Sci Transl Med 2016;8. 366ra165 LP-366ra165.

[57] Bandodkar AJ, et al. Battery-free, skin-interfaced microfluidic/electronic systems for simultaneous electrochemical, colorimetric, and volumetric analysis of sweat. Sci. Adv. 2019;5:eaav3294.

[58] Reeder JT, et al. Waterproof, electronics-enabled, epidermal microfluidic devices for sweat collection, biomarker analysis, and thermography in aquatic settings. Sci Adv 2019;5:eaau6356.

[59] Godek SF, Bartolozzi AR, Godek JJ. Sweat rate and fluid turnover in American football players compared with runners in a hot and humid environment. Br J Sports Med 2005;39:205−11.

[60] Ogai K, Fukuoka M, Kitamura K-I, Uchide K, Nemoto T. A detailed protocol for perspiration monitoring using a novel, small, wireless device. J Vis Exp 2016; 54837. https://doi.org/10.3791/54837.

[61] Scott M, Sawka M, Wenger C. Hyponatremia associated with exercise: risk factors and pathogenesis. Exerc Sport Sci Rev 2001;29:113−7.

[62] Mishra A, Greaves R, Massie J. The relevance of sweat testing for the diagnosis of cystic fibrosis in the genomic era. Clin Biochem Rev 2005;26:135−53.

[63] Biagi S, Ghimenti S, Onor M, Bramanti E. Simultaneous determination of lactate and pyruvate in human sweat using reversed-phase high-performance liquid chromatography: a noninvasive approach. Biomed Chromatogr 2012;26:1408−15.

[64] Jia W, et al. Electrochemical tattoo biosensors for real-time noninvasive lactate monitoring in human perspiration. Anal Chem 2013;85:6553−60.

[65] Bruen D, Delaney C, Florea L, Diamond D. Glucose sensing for diabetes monitoring: recent developments. Sensors 2017;17:1866.

[66] Abellán-Llobregat A, et al. A stretchable and screen-printed electrochemical sensor for glucose determination in human perspiration. Biosens Bioelectron 2017;91:885−91.

[67] Harker M, Coulson H, Fairweather I, Taylor D, Daykin CA. Study of metabolite composition of eccrine sweat from healthy male and female human subjects by 1H NMR spectroscopy. Metabolomics 2006;2:105−12.

[68] Boyne MS, Silver DM, Kaplan J, Saudek CD. Timing of changes in interstitial and venous blood glucose measured with a continuous subcutaneous glucose sensor. Diabetes 2003;52. 2790 LP-2794.

[69] Lee H, et al. A graphene-based electrochemical device with thermoresponsive microneedles for diabetes monitoring and therapy. Nat Nanotechnol 2016;11:566.

[70] Girard P, Beraud A, Sirvent A. Study of three complementary techniques for measuring cutaneous hydration in vivo in human subjects: NMR spectroscopy, transient thermal transfer and corneometry - application to xerotic skin and cosmetics. Skin Res Technol 2000;6.

[71] Alanen E, Nuutinen J, Nicklén K, Lahtinen T, Mönkkönen J. Measurement of hydration in the stratum corneum with the MoistureMeter and comparison with the Corneometer. Skin Res Technol 2004;10.

[72] Madhvapathy SR, et al. Epidermal electronic systems for measuring the thermal properties of human skin at depths of up to several millimeters. Adv Funct Mater 2018;28: 1802083.

[73] Krishnan S, et al. Multimodal epidermal devices for hydration monitoring. Microsystems Nanoeng 2017;3:17014.

[74] Crawford KE, et al. Advanced approaches for quantitative characterization of thermal transport properties in soft materials using thin, conformable resistive sensors. Extrem Mech Lett 2018;22:27−35.

[74a] James CA, Richardson AJ, Watt PW, Maxwell NS. Reliability and validity of skin temperature measurement by telemetry thermistors and a thermal camera during exercise in the heat. J Therm Biol 2014;45:141−9.

[75] Hattori Y, et al. Multifunctional skin-like electronics for quantitative, clinical monitoring of cutaneous wound healing. Adv Healthc Mater 2014;3:1597−607.

[76] Webb RC, et al. Ultrathin conformal devices for precise and continuous thermal characterization of human skin. Nat Mater 2013;12:938.

[77] Krishnan SR, et al. Wireless, battery-free epidermal electronics for continuous, quantitative, multimodal thermal characterization of skin. Small 2018;14:1803192.

[78] Wang S, et al. Mechanics of epidermal electronics. J Appl Mech 2012;79:31022−6.

[79] Rogers A, Ghaffari JR, Kim D-H. Stretchable bioelectronics for medical devices and systems. 2016. https://doi.org/10.1007/978-3-319-28694-5.

[80] Lu N, Yoon J, Suo Z. Delamination of stiff islands patterned on stretchable substrates. Int J Mater Res 2007;98:717−22.

[81] Shin G, et al. Flexible near-field wireless optoelectronics as subdermal implants for broad applications in optogenetics. Neuron 2017;93:509−21. e3.

[82] Verdier-Sévrain S, Bonté F. Skin hydration: a review on its molecular mechanisms. J Cosmet Dermatol 2007;6:75−82.

[83] Weinbaum S, Xu LX, Zhu L, Ekpene A. A new fundamental bioheat equation for muscle tissue: Part I—blood perfusion term. J Biomech Eng 1997;119:278−88.

[84] REGAS FC, EHRLICH HP. Elucidating the vascular response to burns with a new rat model. J. Trauma Acute Care Surg. 1992;32.

[85] Krishnan SR, et al. Epidermal electronics: wireless, battery-free epidermal electronics for continuous, quantitative, multimodal thermal characterization of skin (small 47/2018). Small 2018;14:1870226.

[86] Koral K, Blackburn T, Bailey AA, Koral KM, Anderson J. Strengthening the argument for rapid brain MR imaging: estimation of reduction in lifetime attributable risk of developing fatal cancer in children with shunted hydrocephalus by instituting a rapid brain MR imaging protocol in lieu of head CT. Am J Neuroradiol 2012;33. 1851 LP-1854.

[87] Krishnamurthy S, Schmidt B, Tichenor MD. Radiation risk due to shunted hydrocephalus and the role of MR imaging−safe programmable valves. Am J Neuroradiol 2013; 34. 695 LP-697.

[88] Krishnan SR, et al. Epidermal electronics for noninvasive, wireless, quantitative assessment of ventricular shunt function in patients with hydrocephalus. Sci Transl Med 2018;10:eaat8437.

[89] Matsumoto M, et al. Estimating the cost of skin cancer detection by dermatology providers in a large health care system. J Am Acad Dermatol 2018;78:701−9. e1.

[90] Panther DJ, Jacob SE. The importance of acidification in atopic eczema: an underexplored avenue for treatment. J Clin Med 2015;4:970−8.

[91] Adams JL, et al. Multiple wearable sensors in Parkinson and Huntington disease individuals: a pilot study in clinic and at home. Digit. Biomarkers 2017;1:52−63.

Food safety applications

Daniel Dooyum Uyeh, PhD [1], **Wonjin Shin**[1], **Yushin Ha, PhD** [2], **Tusan Park, PhD** [2]

[1]*Department of Bio-Industrial Machinery Engineering, Kyungpook National University, Daegu, Republic of Korea;* [2]*Professor, Bio-Industrial Machinery Engineering, Kyungpook National University, Daegu, Republic of Korea*

1. Introduction

According to the World Health Organization (WHO), food- and waterborne diseases are caused mostly by ingestion of infectious pathogens or toxic chemicals [1]. Over 250 foodborne diseases and a wide variety of associated symptoms (food poisoning) have been documented [2]. Food poisoning is a severe gastrointestinal condition triggered by toxic chemicals and pathogens in food. Although rarely serious, food poisoning can be life threatening if left untreated or recovery is not achieved within the expected time [3]. Usually, symptoms can be observed within a few hours or days of ingesting potentially harmful substance. WHO listed the common bacterial agents in food poisoning, to include *Listeria monocytogenes*, *Vibrio cholerae*, *Salmonella* spp., *Campylobacter* spp., and enterohemorrhagic *Escherichia coli*, as well as viral pathogens such as Norovirus and Hepatitis A. Furthermore, fish- and waterborne trematodes, *Ascaris lumbricoides*, *Cryptosporidium* spp., *Giardia* spp., *Entamoeba histolytica*, *Echinococcus* spp. and *Taenia solium* are common pathogenic parasites that can also cause foodborne illness. Other causative agents include proteinaceous infectious particles (prions) and toxic chemicals such as persistent organic pollutants, heavy metals, and naturally occurring toxins like marine biotoxins.

Approximately 1 in 10 people fall ill after eating contaminated food around the world and 420,000 die every year, resulting in the loss of 33 million healthy lives annually [4]. Children under 5 years old make up 40% of the people affected by foodborne poisoning, resulting in 125,000 deaths annually. Consumption of contaminated food often leads to diarrheal diseases, which cause 550 million people to fall ill and 230,000 to die every year. There is an inseparable linkage between food safety, nutrition, and food security. A vicious cycle of disease and malnutrition that affects mostly sick people, infants, young children, and the elderly is created. Children younger than 5 years, adults older than 65 years, pregnant women, and immunocompromised patients with diabetes, cancer, and HIV infection are the major groups at risk of food poisoning [1]. Access to potable water and lack of proper storage facilities are the major reasons for food contamination, especially in developing countries.

Various mass food poisoning cases have been reported worldwide with high mortality rates of up to millions. For example, *Listeria* was identified from meat in South Africa in 2018; several insecticides were found from candy in Pakistan in 2016; various strains of *E. coli* were identified from chicken salad, beef, flour, and raw clover sprouts in the United States between 2014 and 2018; Norovirus was found from kimchi in Republic of Korea in 2013. There is a need for rapid methods to identify these poisoning agents, as these foods are consumed in large quantities in these countries. In addition, due to environmental changes and subsequent shortages, global food safety is being challenged from importing and transporting food around the world, leading to possible contamination. Furthermore, new and emerging toxins as well as antibiotic-resistant bacteria further complicate food safety. Countries and organizations around the world have established a food safety plan to reduce food poisoning, for example, a private organization, The Consumer Goods Forum, organized by executives and stakeholders in food industry from over 70 countries, works to ensure global food safety through establishing the Global Food Safety Initiative. Many countries, for example, countries in the European Union, Japan, Canada, United States, Republic of Korea, etc. have implemented special acts on safety management to prevent foodborne poisoning from imported foods [5], mandating preregistration of import facilities and empowering suspension of import as necessary.

This chapter examines the sources of food poisoning, the regulations on foods, the standard methods used in their identification, and quantification. Furthermore, this chapter discusses the advances made in the quantification of foodborne pathogens using portable electronic devices with a focus on smartphone.

2. Major foodborne outbreaks around the world

2.1 Biological food contaminants

In this section, the common agents of foodborne poisoning in the world are discussed. They are categorized into parasites, bacteria, and viruses.

2.1.1 Parasites

Ascariasis is a disease caused by *Ascaris lumbricoides* and estimated to have infected about 0.81−1.2 billion people in the world [2]. Whipworm, hookworm, and *A. lumbricoides* are parasitic worms known as soil-transmitted helminths. The major burdens of parasitic disease in the world are caused by these three parasites. Human infection mostly results from ingestion of the eggs by consuming fruits or vegetables contaminated with these parasites with little care in washing, peeling, or cooking. People infected with *Ascaris* spp. are often asymptomatic. Even in symptomatic situations, only light abdominal discomfort is felt. Severe infections can lead to blockage in the intestinal and consequently impairment in the growth of children.

Anisakiasis is caused by anisakid nematodes through ingestion of infected, raw, or undercooked fish, as shown in Fig. 11.1 [2]. **Amebiasis** is caused by *Entamoeba histolytica*, a disease that can affect anyone, although commonly found to affect people in tropical areas with poor sanitary conditions. Microscopic examinations of anisakiasis and amebiasis are difficult because of their similarities to other parasites. Infected people do not always become sick [2].

2.1.2 Bacteria

Botulism is caused by *Clostridium botulinum*, which is rare but serious foodborne illness. It produces a toxin that attacks the body's nerves and is found mostly in canned foods. Weakness of muscles is initially observed on face, eyes, throat, and mouth, which spreads to torso, neck, arms, and legs. Other symptoms include weakening of the breathing muscles that can lead to breathing difficulty and even death [2]. The U.S. Food and Drug Administration (FDA) [6] has set safe limits to its presence in 25 g sample of ready-to-eat foods.

Cholera is caused by *Vibrio cholerae* and can be life threatening. It can be found in a variety of foods such as rice, millet, meat, dairy products, and fruits. Despite it can be life threatening, prevention and treatment are possible. Its main mode of infection is usually water used in cooking and for drinking [2]. Presence of toxigenic O1 and O139, or non-O1 and non-O139 in 25 g sample of ready-to-eat food is the safe limit set by the US FDA [6].

Escherichia coli forms the normal bioflora in the intestines of humans and animals. There are different species of *E. coli* that are harmless and makes up an important part of a healthy human intestinal tract. Nevertheless, some species are pathogenic, mostly causing diarrhea. Diarrhea from *E. coli* results from ingesting contaminated water or food, or animal–human contact. Enterohemorrhagic *E. coli* or verocytotoxin-producing *E. coli* is the most common Shiga toxin producing *E. coli* related to foodborne outbreaks [2]. The International Commission on Microbiological Specifications for Foods [7] set the microbiological limits for frozen fruits and vegetables to be 10^2 CFU/g for the *E. coli*.

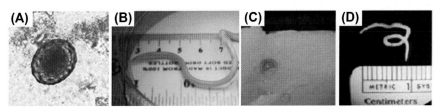

FIGURE 11.1

Image of fertilized eggs of *A. lumbricoides* in unstained wet mounts of stool (A); adult female *A. lumbricoides* (B); coiled worms of anisakid nematode in fresh, nonfrozen cod fillet (C); and larva of an anisakid nematode (D).

Adopted from DPDx, Orange County Public Health Laboratory, Santa Ana, CA, USA. Available from: https://www.
cdc.gov/foodsafety/diseases/#viral.

Salmonellosis, which is caused by *Salmonella* spp., is a major cause of bacterial enteric illness in humans and animals. An estimated 1.4 million cases of salmonellosis occur each year in humans in the United States [8]. They are commonly found in poultry products and milk but can also be found in vegetables and fruits. New Zealand and Australian standards set the limits for Salmonella in grains and cultured seeds to zero [7].

Clostridium perfringens is commonly found in various environmental sources and are a spore-forming gram-positive bacterium. They can also be found in the intestines of humans, animals, on raw meat and poultry as well as dried or precooked foods. *C. perfringens* is mostly anaerobic but can also grow with the presence of little oxygen, multiplying rapidly under such condition. Foods prepared in large quantities and set aside for a long time in a warm condition is the common route of infection. Symptoms with *C. perfringens* infection include abdominal cramps and diarrhea typically within 8–12 h lasting for less than a day. Human-to-human transmission does not occur with *C. perfringens*.

Campylobacter spp. are found in raw or undercooked poultry, as well as contaminated water, contact with animals, or unpasteurized milk [2].

Listeriosis is a serious infection usually caused by eating food contaminated with *Listeria monocytogenes*. The infection is most likely to sicken pregnant women and their newborns, adults aged 65 or older, and people with compromised immune systems [2]. Ireland set standards for ready-to-eat foods to less than 20 CFU/g for *Listeria* spp. [5,7].

2.1.3 Viruses

Hepatitis A is caused by Hepatitis A virus, which is a communicable disease affecting the liver. It is preventable using the vaccine. The fecal-oral route or consumption of contaminated food or water are the main route of person-to-person transmission. Hepatitis A is a self-limited disease that does not result in chronic infection. Symptoms include fatigue, low appetite, stomach pain, nausea, and jaundice, which usually resolve within 2 months of infection. Most children less than 6 years old have an unrecognized infection or are asymptomatic. Antibodies produced in response to hepatitis A infection last for life and shield against reinfection [2].

2.2 Chemical food contaminants

Acute foodborne illness is also caused by chemical hazards usually from ingestion of abnormally high doses of chemicals. Chemical toxins can come from natural sources such as those produced by plants, animals, or microorganisms, for example, scombrotoxin in fish and aflatoxins in peanuts. Other sources of poisoning include intentionally added food chemicals (for example, sodium nitrate) beyond the acceptable limits established by regulating agencies. The European Food Safety Authority sets the limit for sodium nitrate to 3.7 mg per kg of body weight per day [9]. Furthermore, added chemicals used in sanitation or hygienic maintenance can accidentally contaminate food during processing. Risk factors from chemicals include the quantity being exposed to, and the toxicity of the chemical.

2.3 Food allergens

Another aspect of foodborne illness is food allergy that can be sometimes life threatening. These are the substances in food that can cause a dangerous reaction in people who are allergic. Despite efforts using an enacted "the food consumer protection act" to ensure labeling of prepackaged food with allergen-related ingredients, there still are hidden amounts in processed food from possible cross-contamination occurring in the processing and transportation of food samples. Most cases of food allergy are found in developed countries [10]. Common allergens include peanuts, fish, and dairy products.

3. Laboratory methods to detect foodborne poisoning
3.1 Biological methods

The standard biological methods for detecting foodborne poisoning include (1) culture-based methods mostly carried out using agar, (2) immunoassays, and (3) polymerase chain reaction (PCR). These methods are summarized in Fig. 11.2.

3.1.1 Culture-based methods

Culture-based methods are the oldest methods used in the detection of microorganisms. These methods are cost-effective with high success rate [11]. However, the time used in culturing and growth of the microorganism can be limiting.

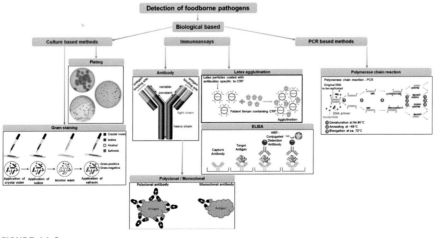

FIGURE 11.2

Biological methods used in detecting foodborne pathogens.

Consequently, this can lead to public health hazards where real-time results are required. Moreover, high expertise is also required.

Highly pathogenic *E. coli* O157:H7 can be cultured on Sorbitol MacConkey agar (SMAC), which has shown a high success rate and cost-effectiveness. This culture method is based on the principle of sorbitol fermentation. Nevertheless, limitations are associated with false-positive results, arising from the long culture time and the difficulty in distinguishing non-O157 and O157 STEC serotypes on sorbitol fermentation. The use of chromogenic substances such as CHROMagar manufactured by DRG International Inc., USA, has proven to be comparatively more effective than SMAC. Despite the comparative advantage of using CHROMagar, it still has a limitation in sensitivity to some strains.

Cefsulodin−Irgasan−Novobiocin agar is a selective medium known for better discrimination between bacterial species, which was used to differentiate *Yersinia enterocolitica* and non-*Y. enterocolitica*. The chromogenic medium (resulting in purple/blue colonies) for *Y. enterocolitica* is used where agar has fermentable sugar cellobiose, a chromogenic substrate, and selective inhibitor, which suppress the competing bacteria. Both *Y. enterocolitica* and *Y. pseudotuberculosis* are important food pathogens causing yersiniosis. This method was used to study the contamination of tofu [12].

Many microorganisms tend to enter a starvation mode of metabolism under stress conditions. However, they remain viable but nonculturable (VBNC), which cannot be grown on conventional culture media but can signal pathogenic pathways. Detection of these pathogens is a major challenge for food safety. As no colonies are formed, other methods such as fluorescent dyes are used for the detection of bacteria that are VBNC. Acridine orange dye can be used on these pathogens to identify their viability, where it binds to the DNA of bacteria. Actively reproducing cells appear green whereas slow-growing or nonreproducing cells appear orange. Another dye used to detect VBNC bacteria is fluorescein isothiocyanate. The principle is to detect the enzyme activity of living cells. If there is the presence of any living cells, violet or blue color is seen.

Potable water, pasteurized milk, and processed food are vulnerable to culturable bacteria. Some of the foodborne pathogens that fall under this category include *Campylobacter jejuni*, *Enterobacter aerogenes*, *Enterococcus faecalis*, *E. coli*, *Pseudomonas aeruginosa*, *Salmonella* Typhimurium, *Shigella dysenteriae*, *Shigella Sonnei*, and *V. cholerae*.

3.1.2 Immunoassays

Immunoassays are standard laboratory methods used in identifying foodborne pathogens. They are rapid, relatively easier to perform compared to the culture-based methods, and cost-effective. Immunoassays are typically performed as prescreening before exploring PCR-based methods [11]. Enzyme-linked immunosorbent assay (ELISA) is one of the most used immunoassays. Purity and antibody specificity are the main factors that affect immunoassay performance. Two types of antibodies can be used: polyclonal antibody that can bind to multiple epitopes in the target, and monoclonal antibody to a single type of epitope. Consequently, low specificity and sensitivity are expected with polyclonal antibodies.

The major advantage of this method is the specific binding to the target and development of visually identifiable coloration. However, false-positive results can be obtained through nonspecific binding to the incorrect target. False positives can also be recorded due to the contamination in the sample. Coloration is achieved by conjugating chemicals or enzyme substrate to the assay. This includes the reaction between bovine serum albumin and 2,2-azino-bis (3-ethylbenzthiazoline-6-sulphonic acid) prepared in 0.05 M phosphate-citrate buffer. Another example is horse radish peroxidase (HRP—enzyme) with tetramethylbenzidine (substrate) with gradual development in coloration. This method was used for the sandwich ELISA detecting *Listeria* sp. Another most commonly used substrate is p-nitrophenyl phosphate (pNPP) that binds specifically to alkaline phosphatase (ALP). In one of the experiments [13], pNPP was used as the substrate for detecting *E. coli* O157:H7.

Engineering of bispecific antibodies was demonstrated to recognize both the red blood cells (RBCs) of human and *L. monocytogenes,* a common foodborne pathogen [14]. Initially, a mixture of anti-RBC and anti-*Listeria* antibodies were added, followed by gradual reoxidation of the reduced disulfides. This facilitates association of the separated antibody chains and formation of hybrid immunoglobulins with an affinity for *L. monocytogenes* and human RBC. The bispecific antibodies caused the agglutination of the RBCs only in the presence of *L. monocytogenes* cells. The agglutination process showed red colored clumps in the presence of *L. monocytogenes* and was readily visible to the naked eye. This was found to be a simple approach for the rapid and highly specific screening of various pathogens in their biological niches.

Monoclonal antibodies are preferred over polyclonal antibodies as addressed earlier. Although sensitivity and specificity are its major positive features, production of monoclonal antibody is a laborious process and is not cost-effective.

ELISA methods have been improved to suit the ever-emerging new applications. Blocking ELISA was designed for targeting *E. coli* O157:H7 lipopolysaccharide as antigen [15]. These were successful in detecting the pathogen in cattle and were found to be more sensitive than the normal ELISA. Indirect ELISA has been used for detecting anti-O157 antibodies in the serum of cattle as well as humans. However, there still exist chances of false-positive due to cross-reactivity of antibodies.

3.1.3 Polymerase chain reaction-based methods

The principle behind PCR is the amplification of genes of pathogens for more accurate and sensitive analysis. Specific primers are developed for each gene. Agarose gel electrophoresis and subsequent staining with ethidium bromide are used for the identification of PCR products. Various types of PCR assays have emerged, providing very promising methods for identifying and quantifying pathogens. The sensitivity and rapid output are the major advantages of PCR. It is faster when compared to culture-based methods and comparable to immunoassays. The amplified product can be obtained as early as 30 min, and although the total assay time would be at least 2 h. Multiple sets of primers can be used to detect multiple targets simultaneously. Detection limit can be as low as femtograms (10^{-15} g). However, despite the advancement in PCR technology, there are many difficulties in these methods, leading to inconsistent results: (1) they are not cost-effective; (2) substantial time and labor are necessary for cell lysis and nucleic acid extraction; (3) cross-contamination and subsequent failure in amplification due to the presence of inhibitory substance or competing DNA from the nontarget cells. Furthermore, PCR methods are not able to differentiate between live and dead cells. The primary disadvantage of all PCR methods is the chances of false-positive signal due to nonspecific amplification.

3.2 Chemical detection

Chromatographic methods, including high-performance liquid chromatography (HPLC) and thin-layer chromatography, are popular throughout the world for food analysis and its quality control. In HPLC, through a column filled with absorbent material, pressurized solvent is passed in liquid form that contains a sample. Each molecule in a sample interacts differently with the adsorbent material. The flow rate of each molecule is varied, leading to separation. These methods have been applied to detecting and quantifying intentional additives, adulterants, and contaminants to food. Liquid chromatograph have also been widely used to assess the composition of food samples, such as protein quality through analyzing amino acids, quality and fat adulteration through analyzing lipids and fatty acids, beverage quality through analyzing sugar contents, storage stability through analyzing biogenic amines, addition of nutrients, colorants, antioxidants, and preservatives through analyzing inorganic acids. Changes in texture and aroma of the food can also be monitored for detecting foodborne poisons. Fig. 11.3 summarizes the chemical detection methods.

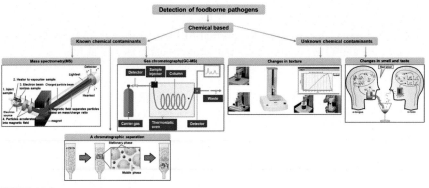

FIGURE 11.3

Chemical methods used in detecting foodborne poisons.

4. Smartphone-based assay for food safety

There have been advances in the use of portable devices especially smartphones in identifying contaminants that cause food poisoning [16]. Most of the developed devices adopt the colorimetric approach using the smartphone's camera, where transmittance, absorbance, or reflectance is used in quantifying the target.

The camera of the smartphone is the component that captures the image and subsequently average light intensity from the area of interest. Such detection can further be enhanced by adding an additional lens and fabricating additional components. Table 11.1 shows a summary of smartphone-based monitoring of food safety.

4.1 Allergen detection using smartphone-based colorimetry

Coskun et al. [10] fabricated a compact *iTube* that was attached to the camera unit of a smartphone, and demonstrated a colorimetric assay using a food allergy test kit specific to peanuts. The target food sample was ground into fine particle size, mixed with hot water, and added with a solvent for extraction. Three drops of the sample solution and the control solution without food content were added separately to two different tubes. The test and control tubes were rinsed sequentially with three drops of blue-labeled conjugate, green-labeled substrate, and red-labeled stop solution after incubation. A wash buffer was used to clean the tubes in between each step. The resultant blue and red mixture color activated in the tubes was measured using a digital reader, which utilized Galaxy S II, an Android phone (8 MP camera with F/2.65 aperture and 4 mm focal length lens). The test and control tubes were inserted from the side and were vertically illuminated by two separate light-emitting diodes (LEDs). The illuminated light was absorbed by the allergen assay activated within the tubes, which caused the intensity of the light to change. This change in light was acquired with the images captured by a smartphone. They specifically selected the wavelength of these LEDs (650 nm) to match the absorption spectrum of the colorimetric assay performed in the test tube. Two diffusers were inserted between the LEDs and the tubes to uniformly illuminate the cross-section of each tube. The transmitted light through each tube of interest was collected through two circular apertures of 1.5 mm diameter to be imaged onto the digital camera of the smartphone using a planoconvex lens. The acquired pictures of tubes for control and sample were converted into binary masked images. A 300×300 pixels frame was used to calculate the transmission signal. The resulting signals were divided by a normalization factor and the normalized signals from test and control tubes were used to determine the relative absorbance (A) of the assay, where $A = \log_{10} I_{control}/I_{tube}$. This absorbance was then plotted against the known allergen concentration (0–25 ppm) within the sample to construct a standard curve. Using the slope of this standard curve, the final concentration of the allergen could be evaluated. They reported that transmission images of the sample and control tubes were digitally processed within 1 s using a smart application running on the smartphone. If the optical properties (e.g., reflection, absorption) of the tube containers were the same for both the sample and control tubes and that the

Table 11.1 Use of smartphone for monitoring food safety.

Device	Principle	Sample	Target	Linear range	Limit of detection	References
Smartphone	ELISA	Cookies	Allergen	1–25 ppm	~1 ppm	[10]
Field-portable fluorescent imager + Smartphone	ELISA	Egg Yoghurt	E. coli O157:H7	10^0–10^6 CFU/mL	~10 CFU/mL	[17]
LFIS reading system + Smartphone	Lateral flow immunoassay	Tap water	Cadmium ion	0.1953–50 ng/mL	0.16 ng/mL	[18]
		Swine urine	Clenbuterol	0.0625–1 ng/mL	0.046 ng/mL	
		Swine fecal	Porcine epidemic diarrhea virus	0.078–20 μg/mL	0.055 μg/mL	
Smartphone	Biochemical principle using enzymes	Spice mixtures and bouillon	(L)-glutamate	0.5–5.0 mmol/L	0.028 mmol/L	[19]
Smartphone	Chromatography	Milk	Tetracycline	0.5–10 μg/mL	0.5 μg/mL	[20]
Smartphone	Immunoassay	Milk	Antirecombinant bovine somatotropin	–	–	[21]
Smartphone	Immunoassay	Milk	Alkaline phosphatase	0.1–150 U/L	0.1 U/L	[22]
Paper-based analytical device + smartphone	Electrokinetic stacking	Reagent	Sudan III	5–40 μM	5.2 μM	[23]
Electrochemical sensor + Smartphone	Aptamer–molecule-specific reaction + enzyme	Milk	Kanamycin	(If no enzyme) 50–150 ng/mL 0.03–20 pg/mL, 20–150 pg/mL	30.0 fg/mL	[24]
Smartphone + EOC-CIS biosensor	Immunoassay	Sea food	Vibrio parahaemolyticus	1×10^4 –2.5×10^6 CFU/mL	1.4×10^4 CFU/mL	[25]

Detection platform	Detection method	Sample	Analyte	Linear range	LOD	Reference
Smartphone + colorimeter	Euclidean distance	Water	Iron	0.1–1.0 ppm	0.1 ppm	[26]
Smartphone	Chemical reaction	River water	Mercury ion	5–250 µM	80 nM	[27]
Smartphone	Chemical reaction	Fruits	Calcium carbide			[28]
Electrochemical sensor + smartphone	ELISA	Egg white	Ovalbumin	–	0.003 mg/kg	[29]
		Hazelnut,	Cor a1	–	0.089 mg/kg	
		Milk	Casein,	–	0.170 mg/kg	
		Peanut	Ara h1	–	0.007 mg/kg	
		Wheat	Gliadin	–	0.075 mg/kg	
Smartphone	Lateral flow aptamer assay	Tap water	Mercury ions	$10–10^4$ ppb	5 ppb	[30]
			Ochratoxin A	0.01–50 µg/mL	3 ng/mL	
			Salmonella ATCC 50761	150–2000 CFU/mL	85 CFU/mL	
Paper-based microfluidic + Smartphone	Chemical reaction	Milk	Melamine	0.1–1000 ppm	0.1 ppm	[31]
Smartphone	Chemical reaction	Water	Fluoride	0–2 mg/L	–	[32]
Bionic electronic eye + Smartphone	Cell-based assay	Shellfish	Okadaic acid	5–200 µg/L	3.4083 µg/L (using HepG2) 13.4456 µg/L (using THP-1)	[33]
Smartphone	PtBA assay aggregation	–	E. coli O157:H7 / Salmonella enterica	YES/NO YES/NO		[34]
Paper-based microfluidic + Smartphone	Chemical reaction	Red wines	Flavor	–	–	[35]
LFIA reader + Smartphone	Lateral flow immunoassay	Maize	Aflatoxin B1	5–1000 µg/kg	5 µg/kg	[36]
Smartphone	Chemical reaction	Water	pH	1–12	–	[37]

Continued

Table 11.1 Use of smartphone for monitoring food safety.—cont'd

Device	Principle	Sample	Target	Linear range	Limit of detection	References
Smartphone	Absorbance	Vitamin C supplement, orange juice	Ascorbic acid	20–80 µg/mL	5 µg/mL	[38]
Electrochemical sensor + smartphone	ELISA	Standard clenbuterol sample	Clenbuterol	0.3–100 ng/mL	0.076 ng/mL	[39]
Smartphone	Chemical reaction	Fat-free milk	E. coli O157:H7	–	10 CFU/mL	[40]
Smartphone	Mie scattering	Ground beef	E. coli K12	10^1–10^8 CFU/mL	10 CFU/mL	[41]
Smartphone	Chemical reaction	Water	Formaldehyde	0.2–2.5 ppm	–	[42]
Smartphone	Simultaneous reaction and electrokinetic stacking	Water	Copper ion	0.1–10 mM	30 µM	[43]
Smartphone	Absorbance	Mineral water	Oxygen pH	7.8–8.8 mg/L O_2 7.1–7.5 pH		[44]
		Rainwater		3.7–4.7 mg/L O_2 5.8–6.2 pH		
		Pure water		6.6–7.6 mg/L O_2 6.6–7.0 pH		
		Lake water		3.1–4.1 mg/L O_2 7.2–7.6 pH		
Smartphone	ELISA	Milk Cheese Water	Salmonella enteritidis	94.2–101.2 CFU/mL		[45]
Smartphone	Luminescence resonance energy transfer	Spiked apple juice	Thiram	0.1 µM–1 mM	0.1 µM	[46]

Platform	Method	Sample	Analyte	Linear range	LOD	Ref.
Smartphone	Immunoassay microarray	Milk	Tetracyclines	0.2–3.2 ng/mL	–	[47]
Lab-on-a-chip device + Smartphone	Competitive ELISA	–	Quinolones BDE-47	0.15–3.6 ng/mL 10^{-3}–10^4 µg/L	–	[47]
Smartphone	Chemical reaction	Minced meat	Isobutylamine Isopentylamine Triethylamine	–	~1 ppm	[48]
Paper-based microfluidic + Smartphone	Immunoassay	Water	Norovirus	10–10^7 pg/mL	10 pg/mL	[49]
Smartphone	Chemical reaction	Water	Bisphenol	–	4.4 µM	[50]
Paper-based microfluidic + Smartphone	Immunoassay	Water	E. coli K12	10–10^5 CFU/mL	10 CFU/mL	[51]
Field-deployable handheld device	Immunoassay	Iceberg lettuce	E. coli K12 E. coli O157:H7	10–10^7 CFU/mL 10–10^4 CFU/mL	10 CFU/mL	[52]
Smartphone	Immunoassay	Water	Cr (VI) Total chlorine Caffeine E. coli K12	0.05–50 ppm 1.5–5 ppm 100–400 ppm 10–10^5 CFU/mL	2.5 ppm 1.5 ppm 100 ppm 10 CFU/mL	[53]

illumination is uniform, that is, approximately the same for both tubes, this was correlated to the concentration of the allergen in the sample tube.

4.2 Streptomycin detection using smartphone-based colorimetry of aptamer-conjugated gold nanoparticles

Liu et al. [54] demonstrated colorimetric indicator for quantifying streptomycin, utilizing aptamer-conjugated gold nanoparticles (AuNPs), a battery-powered optosensing accessory attached to the smartphone's camera. The sensing platform was assembled onto a smartphone for quantification of streptomycin. The smartphone had an 8 MP rear camera. The system consisted of an adaptor, optical tube for a planoconvex lens, and a plastic diffuser. Other parts were a holder with a black lid used for positioning cuvette, a three-color LED and a battery. An aisle with a diameter of 13 mm was designed as a light path. Black photopolymer resin material was utilized to minimize light leakage. Fig. 11.4 illustrates the device.

This assay is based on the distance-dependent optical property of AuNPs, shifting its color from red toward blue upon aggregation due to their size-dependent surface plasmon resonance absorption. The presence of streptomycin in the sample resulted in their binding to aptamer-conjugated AuNPs and subsequently aggregation. Fig. 11.5 shows the mechanism. The turnover process from dispersion to the aggregation state of AuNPs is associated with the characteristic absorption peak shift from ~ 520 to ~ 625 nm. They thought the increase of streptomycin concentration could result in the systematic increase of the absorbance at 625 nm and the contrary decrease of the absorbance at 520 nm. Subsequently, they recorded the ratio of the absorbance at 625 nm to that at 520 nm could be used for streptomycin quantitation using a smartphone attached with an absorption spectrometer through dual wavelength illuminations.

A custom software application was developed to extract the RGB intensities from the images of smartphone, as shown in Fig. 11.6. This was done by segmenting the images to define the region of interest. Afterward, the extraction of the RGB value of each pixel was carried out, the average transmitted intensity of the three images was calculated, and the absorbance was estimated. The application was further developed to relate these absorbance values to streptomycin concentration using a predetermined standard curve, and to display and wirelessly share the result. The system was evaluated with honey, milk, and tap water.

4.3 Lateral flow aptamer assay using smartphone and upconversion nanoparticles

A lateral flow aptamer assay integrated with a smartphone-based portable device was developed for simultaneous detection of multiple targets using upconversion nanoparticles (UCNPs), as shown in Fig. 11.7. Jin et al. [30] synthesized UCNPs toward enhanced fluorescence intensity, through modifying their surface using a ligand exchange process. The resulting UCNPs were used for lateral flow aptamer

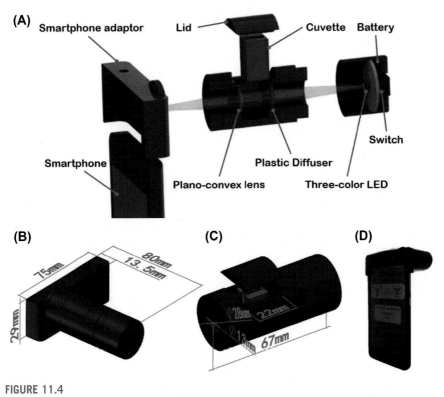

FIGURE 11.4

Three-dimensional printed smartphone-based platform (SBP). Schematic overview of the internal structure of the SBP (A); perspectives of the SBP with dimensions (B–C); SBP installed on a smartphone (D).

Reprinted from Ref. [54] with permission from Elsevier.

assay. *Salmonella* (main analyte), *E. coli*, *Staphylococcus aureus*, and *Bacillus subtilis* were assayed using the developed system. In addition, ochratoxin A was also selected as a target while Aflatoxin B1, kanamycin, and melamine were selected as control groups. For ion target, mercury ion was selected as a target while lead, cupric, and ferrous ion were selected as control groups. Running buffer was prepared followed by adding three UCNPs probes (red, green, and blue) into the buffer. Each colored UCNP probe was conjugated to a different aptamer sequence by a condensation reaction. Subsequently, the as-prepared probes were pipetted separately into a sample solution, not immobilized on a lateral flow assay pad. The lateral flow aptamer assays were submerged into the mixture for 30 min and then removed. The 980 nm laser was used as the light source at a 45 degrees position to the test zone. Complementary ssDNAs to these aptamers were also immobilized to the paper (nitrocellulose) substrate using streptavidin–biotin binding. When target was absent, aptamers hybridized with the surface-bound complementary ssDNA leading

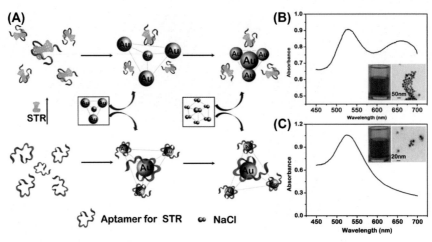

FIGURE 11.5

The sensing mechanism for streptomycin quantitation using aptamer-conjugated AuNPs (A); UV—vis absorption spectrum of AuNP solution (B—C).

Reprinted from Ref. [54] with permission from Elsevier.

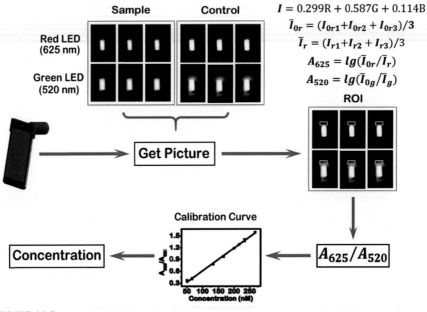

FIGURE 11.6

Image processing steps for streptomycin quantitation.

Reprinted from Ref. [54] with permission from Elsevier.

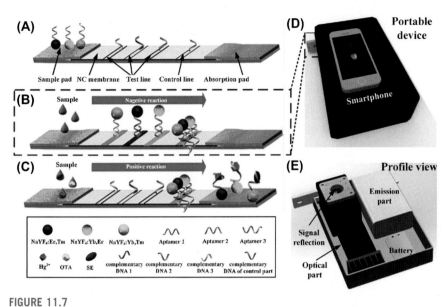

FIGURE 11.7

Schematic of lateral flow aptamer assay for the simultaneous detection of multiple targets.

Reprinted from Ref. [30] with permission from Elsevier.

to the immobilization of UCNPs. However, with target presence (bacteria, metal ions, and small molecules), the aptamers preferentially bound to the corresponding targets leaving fewer hybridization of aptamers and surface immobilization of UCNPs. This reduced the fluorescence intensity from the test zone, which was accordingly quantified using a smartphone-based portable device. Detection ranges were $10-10^4$ ppb, $0.01-50$ μg/mL, and $150-2000$ CFU/mL and detection limits were 5 ppb, 3 ng/mL, and 85 CFU/mL for mercury ion, ochratoxin A, and *Salmonella*, respectively. The system was evaluated in tap water.

4.4 Fluorescence detection of *E. coli* using smartphone camera

Zhu et al. [40] demonstrated the fluorescence detection of *E. coli* captured on the capillary surface. A battery-powered UV LEDs excited the quantum dots on *E. coli,* and the smartphone camera with an additional lens captured the fluorescence emission to quantify the *E. coli* concentration in the sample. The detection limit was $5-10$ CFU/mL in buffer solution. The proof-of-concept was demonstrated using fat-free milk. The capillary tube functioned as a microfluidic channel for liquid delivery and served as a waveguide for the excitation light. These glass capillaries were initially functionalized with anti-*E. coli* O157:H7 antibodies using standard surface chemistry protocols. The fluorescence emission from the quantum dots attached to *E. coli* was imaged and quantified using the microscope attachment to a smartphone. As the guided excitation light propagates perpendicular to the detection path, the optical waveguide scheme functioned as a long-pass filter to remove the scattered UV light. Fluorescent intensity was integrated over a capillary length of ~11 mm. Signals were acquired within 1 s, and subsequently analyzed using ImageJ to quantify *E. coli* concentration in each capillary.

4.5 Pesticide detection on paper using smartphone camera

A paper-based platform integrated with smartphones has been investigated for rapid onsite detection [46]. A 980-nm mini laser, optical filter, and mini cavity were assembled for digitally imaging and quantifying the varied luminescence from paper (Fig. 11.8). Assays were performed without sample purification. Reagents included Yb_2O_3, Y_2O_3, NaOH, NH_4F, Tm_2O_3, $CuCl_2 \cdot 2H_2O$, acetic acid, acetone, and dichloromethane. Targets were thiram, octadecene, oleic acid, polyacrylic acid, nicosulfuron, thifensulfuron, imidacloprid, and 2,4-dichlorophenoxyacetic (2,4-D). Oleate-capped UCNPs were synthesized to detect pesticides. A piece of filter paper was immersed in the prepared aqueous solution to load these UCNPs. A smartphone attachment was 3D printed, consisting of a holder for the mini laser, a dark cavity accommodating an optical filter, and finally a chamber for placing the test paper. A smartphone holder had an optical window with a diameter of 1 cm to allow access

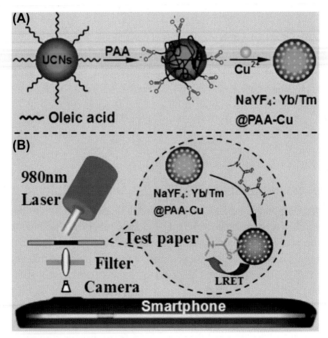

FIGURE 11.8

Smartphone detection of pesticide thiram using upconversion nanoparticles (UCNPs) on paper.

Reprinted from Ref. [46] with permission from Elsevier.

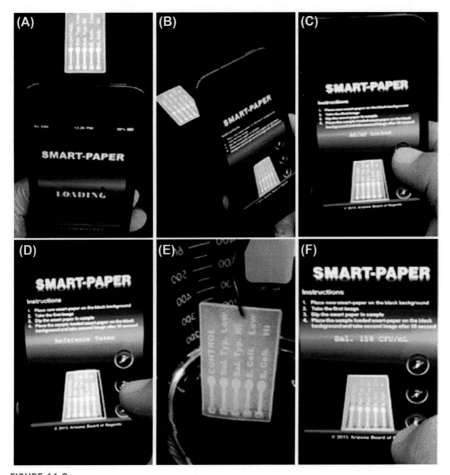

FIGURE 11.9

Smartphone detection of *Salmonella* from a multichannel microfluidic device.

Reprinted from Ref. [55] with permission from Royal Society of Chemistry.

to a smartphone camera. Presence of pesticide, in this case thiram, resulted in quenching blue luminescence through luminescence resonance energy transfer mechanism. The blue channel intensities of images were processed to quantify the thiram concentration through an Android app installed on a smartphone. The detection limit of the system was 0.1 μM.

4.6 *Salmonella* and *E. coli* detection on paper microfluidic devices using smartphone

Paper microfluidic device was used to quantify *Salmonella* [55]. This was carried out by preloading the microfluidic channel with anti-*Salmonella* conjugated submicron particles. Dipping the paper microfluidic device into the solution that contained *Salmonella* resulted in immunoagglutination of these particles within paper fibers. The magnitude of immunoagglutination was measured by collecting the Mie scatter intensities from the images captured by a smartphone at an optimized angle and distance. The method was designed to use a smartphone and a multichannel paper microfluidic device under ambient light. A software application was developed to allow the user to position the smartphone at an optimized angle and distance from the paper microfluidic device and consequently display the bacterial concentration, as shown in Fig. 11.9. The user is asked to position the smartphone to fit the image of the paper microfluidic device into the trapezoid zone on the application screen (A−B). The autofocus/autoexposure is locked by pushing the first button (C) and the reference image is captured (D). Sample is loaded to the paper microfluidic device and the assay result is displayed (F).

Similarly, in Ref. [51], the authors detected *E. coli* from field water samples on paper microfluidics using a smartphone. Multiple channels were printed to cover different ranges of assays, each loaded with varied amount of antibody-conjugated particles. Additionally, contaminants from field water samples were successfully filtered and eliminated through paper fibers.

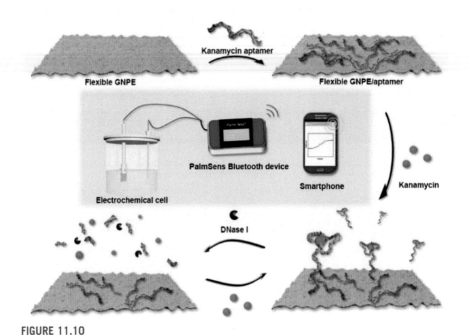

FIGURE 11.10

Potentiometric sensing of kanamycin using aptamers on graphene electrode. Readings are transferred wirelessly to a smartphone.

Reprinted from Ref. [24] with permission from Elsevier.

4.7 Smartphone synchronizes other devices for electrochemical biosensing

A smartphone was connected to a potentiometric sensor wirelessly using Bluetooth [24] toward electrochemical biosensing. A rectangular graphene paper strip was obtained to make the potentiometric sensing electrode. Through noncovalent physical adsorption, graphene paper electrode and the kanamycin aptamer were immobilized on a sensing substrate. The electromotive force between the terminals of a two-electrode system consisted of the graphene paper electrode/aptamer as the working electrode and a saturated commercial Ag/AgCl electrode that served as the reference electrode. Kanamycin (antibiotic drug) was detected via potentiometric detection in real time. Readings from the potentiometric sensor were transferred wirelessly to a smartphone using Bluetooth as shown in Fig. 11.10. This method was evaluated with milk with a detection range of 50–150 ng/mL.

5. Conclusion

We have covered several major sources of foodborne outbreaks around the world. Standard laboratory methods for detecting biological and chemical sources of food poisoning were reviewed, which showed several limitations of false-positive results and their time-consuming nature. Major advances have been made by using smartphones for detecting foodborne pathogens. The most explored principle was the detection of light intensity using the smartphone camera. Smartphone was also connected wirelessly to other devices to record signals from them. There still exist gaps in exploring other components of smartphone toward food safety applications, such as near-field communication, Wi-Fi, and mobile data services.

References

[1] World Health Organization. Foodborne diseases. 2019. Available from: https://www.who.int/topics/foodborne_diseases/en/.

[2] Centers for Disease Control and Prevention. A–Z index for foodborne illness. 2018 [cited 2019 2/1/2019]; Available from: https://www.cdc.gov/foodsafety/diseases/#viral.

[3] Department of Health and Social Care. Food poisoning. 2018. Available from: https://www.nhs.uk/conditions/food-poisoning/.

[4] World Health Organization. Facts on food safety. 2016.

[5] Ministry of Food and Drug Safety. Ministry of food and drug safety reference standards. 2018. Available from: http://www.nifds.go.kr/brd/m_212/down.do?brd_id=board_mfds_304&seq=33094&data_tp=A&file_seq=2.

[6] FDA. Fda and epa safety levels in regulations and guidance [cited 2019 20/2]; Available from: https://www.fda.gov/downloads/food/guidanceregulation/ucm252448.pdf.

[7] Committee on the review of the use of scientific criteria, performance standards for safe food, institute of medicine, nutrition board, board on agriculture, division on earth, life

studies, and national research council, Scientific criteria to ensure safe food. Grove/Atlantic, Inc; 2003.

[8] Brenner FW, Villar RG, Angulo FJ, Tauxe R, Swaminathan B. Salmonella nomenclature. J Clin Microbiol 2000;38(7):2465−7.

[9] The European Food Safety Authority. 3.7 milligrams per kilogram of body weight per day. 2017. Available from: https://www.efsa.europa.eu/en/press/news/170615-0.

[10] Coskun AF, Wong J, Khodadadi D, Nagi R, Tey A, Ozcan A. A personalized food allergen testing platform on a cellphone. Lab Chip 2013;13(4):636−40.

[11] Priyanka B, Patil RK, Dwarakanath S. A review on detection methods used for foodborne pathogens. Indian J Med Res 2016;144(3):327−38.

[12] Weagant SD. A new chromogenic agar medium for detection of potentially virulent yersinia enterocolitica. J Microbiol Methods 2008;72(2):185−90.

[13] Priyanka B, Patil RK, Dwarakanath S. Production of antibodies in chicken for escherichia coli o157: H7. IOSR J Environ Sci Toxicol Food Technol 2012;2(Issue 1):08−15.

[14] Owais M, Kazmi S, Tufail S, Zubair S. An alternative chemical redox method for the production of bispecific antibodies: implication in rapid detection of food borne pathogens. PLoS One 2014;9(3):e91255.

[15] Laegreid W, Hoffman M, James K, Elder R, Kwang J. Development of a blocking enzyme-linked immunosorbent assay for detection of serum antibodies to 0157 antigen of escherichia coli. Clin Diagn Lab Immunol 1998;5(2):242−6.

[16] Kwon O, Park T. Applications of smartphone cameras in agriculture, environment, and food: a review. J Biosyst Eng 2017;42(4):330−8.

[17] Zeinhom MMA, Wang Y, Song Y, Zhu M-J, Lin Y, Du D. A portable smart-phone device for rapid and sensitive detection of E. coli 0157: H7 in yoghurt and egg. Biosens Bioelectron 2018;99:479−85.

[18] Xiao W, Huang C, Xu F, Yan J, Bian H, Fu Q, Xie K, Wang L, Tang Y. A simple and compact smartphone-based device for the quantitative readout of colloidal gold lateral flow immunoassay strips. Sens Actuators B Chem 2018;266:63−70.

[19] Monošík R, Bezerra dos Santos V, Angnes L. A simple paper-strip colorimetric method utilizing dehydrogenase enzymes for analysis of food components. Anal Methods 2015;7(19):8177−84.

[20] Masawat P, Antony H, Namwong A. An iphone-based digital image colorimeter for detecting tetracycline in milk. Food Chem 2015;184:23−9.

[21] Ludwig SKJ, Zhu H, Phillips S, Shiledar A, Feng S, Tseng D, A van Ginkel L, Nielen MWF, Ozcan A. Cellphone-based detection platform for rbst biomarker analysis in milk extracts using a microsphere fluorescence immunoassay. Anal Bioanal Chem 2014;406(27):6857−66.

[22] Yu L, Shi ZZ, Fang C, Zhang YY, Liu YS, Li CM. Disposable lateral flow-through strip for smartphone-camera to quantitatively detect alkaline phosphatase activity in milk. Biosens Bioelectron 2015;69:307−15.

[23] Song Y-Z, Zhang X-X, Liu J-J, Fang F, Wu Z-Y. Electrokinetic stacking of electrically neutral analytes with paper-based analytical device. Talanta 2018;182:247−52.

[24] Yao Y, Jiang C, Jianfeng P. Flexible freestanding graphene paper-based potentiometric enzymatic aptasensor for ultrasensitive wireless detection of kanamycin. Biosens Bioelectron 2019;123:178−84.

[25] Seo S-M, Kim S-W, Jeon J-W, Kim J-H, Kim H-S, Cho J-H, Lee W-H, Paek S-H. Food contamination monitoring via internet of things, exemplified by using pocket-sized immunosensor as terminal unit. Sens Actuators B Chem 2016;233:148−56.

[26] Masawat P, Antony H, Srihirun N, Namwong A. Green determination of total iron in water by digital image colorimetry. Anal Lett 2017;50(1):173—85.

[27] Wang L, Li B, Xu F, Shi X, Feng D, Wei D, Li Y, Feng Y, Wang Y, Jia D. High-yield synthesis of strong photoluminescent N-doped carbon nanodots derived from hydrosoluble chitosan for mercury ion sensing via smartphone app. Biosens Bioelectron 2016;79:1—8.

[28] Maheswaran S, Sathesh S, Priyadharshini P, Vivek B. Identification of artificially ripened fruits using smart phones. In: Intelligent computing and control (I2C2), 2017 International conference on. IEEE; 2017.

[29] Lin H-Y, Huang C-H, Park J, Pathania D, Castro CM, Fasano A, Weissleder R, Lee H. Integrated magneto-chemical sensor for on-site food allergen detection. ACS Nano 2017;11(10):10062—9.

[30] Jin B, Yang Y, He R, Park YI, Lee A, Bai D, Li F, Lu TJ, Xu F, Lin M. Lateral flow aptamer assay integrated smartphone-based portable device for simultaneous detection of multiple targets using upconversion nanoparticles. Sens Actuators B Chem 2018;276:48—56.

[31] Xie L, Zi X, Zeng H, Sun J, Xu L, Chen S. Low-cost fabrication of a paper-based microfluidic using a folded pattern paper. Anal Chim Acta 2018.

[32] Levin S, Krishnan S, Rajkumar S, Halery N, Balkunde P. Monitoring of fluoride in water samples using a smartphone. Sci Total Environ 2016;551:101—7.

[33] Su K, Zhong L, Pan Y, Fang J, Zou Q, Wan Z, Wang P. Novel research on okadaic acid field-based detection using cell viability biosensor and bionic e-eye. Sens Actuators B Chem 2018;256:448—56.

[34] DuVall JA, Julianc CB, Shafagati N, Luzader D, Shukla N, Li J, Kehn-Hall K, Kendall MM, Feldman SH, Landers JP. Optical imaging of paramagnetic bead-DNA aggregation inhibition allows for low copy number detection of infectious pathogens. PLoS One 2015;10(6):e0129830.

[35] Park TS, Baynes C, Cho S-I, Yoon J-Y. Paper microfluidics for red wine tasting. RSC Adv 2014;4(46):24356—62.

[36] Lee S, Kim G, Moon J. Performance improvement of the one-dot lateral flow immunoassay for aflatoxin b1 by using a smartphone-based reading system. Sensors 2013;13(4):5109—16.

[37] Shen L, Hagen JA, Papautsky I. Point-of-care colorimetric detection with a smartphone. Lab Chip 2012;12(21):4240—3.

[38] Aguirre MÁ, Long KD, Canals A, Cunningham BT. Point-of-use detection of ascorbic acid using a spectrometric smartphone-based system. Food Chem 2019;272:141—7.

[39] Dou Y, Jiang Z, Deng W, Su J, Chen S, Song H, Ali A, Zuo X, Song S, Shi J. Portable detection of clenbuterol using a smartphone-based electrochemical biosensor with electric field-driven acceleration. J Electroanal Chem 2016;781:339—44.

[40] Zhu H, Sikora U, Ozcan A. Quantum dot enabled detection of *Escherichia coli* using a cell-phone. Analyst 2012;137(11):2541—4.

[41] Liang P-S, Park TS, Yoon J-Y. Rapid and reagentless detection of microbial contamination within meat utilizing a smartphone-based biosensor. Sci Rep 2014;4:5953.

[42] Guzman JMCC, Tayo LL, Liu C-C, Wang Y-N, Fu L-M. Rapid microfluidic paper-based platform for low concentration formaldehyde detection. Sens Actuators B Chem 2018;255:3623—9.

[43] Liu L, Xie M-R, Fang F, Wu Z-Y. Sensitive colorimetric detection of Cu^{2+} by simultaneous reaction and electrokinetic stacking on a paper-based analytical device. Microchem J 2018;139:357−62.

[44] Xu W, Lu S, Chen Y, Zhao T, Jiang Y, Wang Y, Chen X. Simultaneous color sensing of O_2 and pH using a smartphone. Sens Actuators B Chem 2015;220:326−30.

[45] Zeinhom MMA, Wang Y, Sheng L, Du D, Li L, Zhu M-J, Lin Y. Smart phone based immunosensor coupled with nanoflower signal amplification for rapid detection of *Salmonella enteritidis* in milk, cheese and water. Sens Actuators B Chem 2018;261:75−82.

[46] Mei Q, Jing H, You L, Yisibashaer W, Chen J, Li BN, Zhang Y. Smartphone based visual and quantitative assays on upconversional paper sensor. Biosens Bioelectron 2016; 75:427−32.

[47] Li Z, Li Z, Zhao D, Wen F, Jiang J, Xu D. Smartphone-based visualized microarray detection for multiplexed harmful substances in milk. Biosens Bioelectron 2017;87: 874−80.

[48] Bueno L, Meloni GN, Reddy SM, Paixao TRLC. Use of plastic-based analytical device, smartphone and chemometric tools to discriminate amines. RSC Adv 2015;5(26): 20148−54.

[49] Cho S, Park TS, Reynolds KA, Yoon J-Y. Multi-normalization and interpolation protocol to improve norovirus immunoagglutination assay from paper microfluidics with smartphone detection. SLAS Technology 2017;22(6):609−15.

[50] McCracken KE, Tat T, Paz V, Yoon J-Y. Smartphone-based fluorescence detection of bisphenol a from water samples. RSC Adv 2017;7(15):9237−43.

[51] Park TS, Yoon J-Y. Smartphone detection of *Escherichia coli* from field water samples on paper microfluidics. IEEE Sens J 2015;15(3):1902−7.

[52] You DJ, Geshell KJ, Yoon J-Y. Direct and sensitive detection of foodborne pathogens within fresh produce samples using a field-deployable handheld device. Biosens Bioelectron 2011;28(1):399−406.

[53] McCracken KE, Angus SV, Reynolds KA, Yoon J-Y. Multimodal imaging and lighting bias correction for improved μpad-based water quality monitoring via smartphones. Sci Rep 2016;6:27529.

[54] Liu Z, Zhang Y, Xu S, Zhang H, Tan Y, Ma C, Song R, Jiang L, Yi C. A 3d printed smartphone optosensing platform for point-of-need food safety inspection. Anal Chim Acta 2017;966:81−9.

[55] Park TS, Li W, McCracken KE, Yoon J-Y. Smartphone quantifies salmonella from paper microfluidics. Lab Chip 2013;13(24):4832−40.

Index